Copernicus Books

Sparking Curiosity and Explaining the World

Drawing inspiration from their Renaissance namesake, Copernicus books revolve around scientific curiosity and discovery. Authored by experts from around the world, our books strive to break down barriers and make scientific knowledge more accessible to the public, tackling modern concepts and technologies in a nontechnical and engaging way. Copernicus books are always written with the lay reader in mind, offering introductory forays into different fields to show how the world of science is transforming our daily lives. From astronomy to medicine, business to biology, you will find herein an enriching collection of literature that answers your questions and inspires you to ask even more.

Gian Francesco Giudice

Before the Big Bang

 Springer

Gian Francesco Giudice
Department of Theoretical Physics
CERN
Geneva, Switzerland

Translated by Stephen Lyle

ISSN 2731-8982   ISSN 2731-8990   (electronic)
Copernicus Books
ISBN 978-3-031-69932-0   ISBN 978-3-031-69933-7   (eBook)
https://doi.org/10.1007/978-3-031-69933-7

Translation from the Italian language edition: "Prima del Big Bang" by Gian Francesco Giudice, © Rizzoli 2023. Published by Rizzoli, Mondadori Libri S.p.A., Milano. All Rights Reserved.

This Springer imprint is published by the registered company Springer Nature Switzerland AG
The registered company address is: Gewerbestrasse 11, 6330 Cham, Switzerland

Contents

Prologue

The universe is not only queerer than we suppose,
but queerer than we can suppose.
J.B.S. Haldane

I like traveling by train. I find I can concentrate much better on reading than in the silence of my office. The presence of all those strangers helps me to focus. Perhaps, it's just the way they sit casually beside me and then disappear forever at the next station. Traveling by train is like a journey through life, a metaphor for human beings coming together and then drifting apart. Unfortunately, there are also those annoying individuals who contravene the metaphor, loudly advertising their private lives on the phone, or interrupting earnest readers with small talk. Sometimes, however, there are curious encounters.

Once, on the way back from a physics conference, I was seated on a train reading a paper on quantum cosmology. There were about 50 pages packed with mathematical formulas, and it required all my concentration. At one point, I lifted my eyes from the article and noticed that the little girl on the seat opposite was staring at me with that spontaneous confidence that only children can have with a complete stranger. She couldn't have been more than 10 years old and was accompanied by an elderly woman, perhaps her grandmother.

Taking advantage of my momentary break from reading, she asked me: "What are you reading?" Caught off guard, I found no better answer than: "It's the story of the universe." I smiled at her and lowered my eyes once more to the pages of the article. After a long pause, she replied: "So, if it tells the whole story of the universe, does it say anything about me?". "No, I don't think so," I answered hesitantly, "but I haven't got to the end yet." A moment later, the loudspeaker announced the next station and the elderly lady hurried to put on her coat, pushing the girl toward the exit. I barely lifted my arm, in an embarrassed gesture of farewell.

Left to myself again, I struggled to regain my concentration. If the equations of physics really claim to describe the history of the entire universe, from the Big Bang to today, why don't they say anything about that little girl?

My mind went back to my very first cosmology lecture when I was a student at the University of Padua. I had been so excited by the prospect of learning how the universe was born and how it works today. The lecture got off the mark with barely any philosophical preliminaries. The professor wrote some differential equations on the blackboard and illustrated the possible solutions. He then chose one and pronounced emphatically: "This solution describes our universe."

So, is that all? I wondered with some disappointment. My impression at the time was that cosmology was so superficial as to seem almost comical. In contrast, the question I had been asked only a few minutes earlier seemed to me rather profound. What does it actually mean to give a scientific explanation of the history of the universe? I thought back nostalgically to the many years that had gone by since that first lecture at the university and realized just how much my understanding of physics had changed.

At the next station, other passengers took the seats around me and I quickly returned to my reading, hunched over it like a little creature seeking refuge inside its burrow. My gaze was fixed once again on the sheet when a finger appeared in front of my eyes, pointing at the equations, and a quiet voice said: "What's all that about?" "It's the story of the universe," I answered mechanically.

Darn, it really wasn't my day. It was like living in the film *Groundhog Day*, where every day started in the same way, wiping clean everything that had happened on the previous day. In front of me sat a boy in his late teens. He was wearing a hoodie with the hood pulled down over his head. An earphone dangled from one ear, with the wire running down from one shoulder and disappearing into a pocket of his hoodie. Although I made no attempt to encourage him, the young man was not to be deterred. "I've seen those things on the internet. That's where you can find out about the history of the universe. So, what are all those hieroglyphics?"

I felt a certain remorse for not paying more attention to the little girl, so I placed the article on my lap, settled comfortably in my seat, and began to explain to my travel companion what it means to describe the universe with mathematical equations. The boy sat staring at me in silence, without even blinking. The very stillness of his gaze made me feel somewhat uncomfortable. It was hard to tell whether he was actually listening.

When I had finished, he shook his head and retorted: "I read on the internet that scientists think the universe was created by aliens, and I have the proof. What you call the universe is actually just a canvas with a few LEDs stuck on."

His nonchalance encouraged me to go on and we began to talk about aliens. I suggested experiments that might unmask the tricks used by the extraterrestrials to set up this fictitious reality which could so cleverly beguile the whole of humanity. He dismantled them all, one by one, arguing each time that the aliens were a whole lot smarter than us and that we would never be able to catch them in the act. He didn't have time to explain what evidence he had, but hinted that the CIA in America already had full details on how aliens had taken control of the Earth. By now, the train was slowing down and he had to get off at the next station. He grabbed his smartphone and, after a glance at the screen, informed me that, with all the talk, I had prevented him from finishing his video game. However, he added, he had enjoyed himself anyway. I took it as a compliment.

From the window, I watched him stride away along the station platform, heading who knows where. Then he disappeared forever from my sight, swallowed up by a reality unknown to me. I no longer felt like reading and began absent-mindedly looking out of the window at the landscape as it rushed backward at high speed. Then suddenly, I had an idea: why not actually tell the story of the Big Bang? I quickly turned the pages of the article I had in my hands. The last page had been left blank, and I began to make notes, sketching out a plan.

So, the story I'm about to tell begins there.

1

The Beginning of the Story

I have loved the stars too fondly to be fearful of the night.
Sarah Williams

The talent of a scientist lies in asking the right question. The creative part of scientific research is less about finding the answer than identifying the question. Invention lies in glimpsing some novelty in a question, in looking at reality with new eyes and being drawn toward unexplored territories, then opening up paths never before suspected. It is intuition, rather than knowledge, that triggers this process, leading the scientist to identify the best way to formulate a question and pursue its consequences.

This story, like many other science stories, begins with a question. A question so simple that it could arise spontaneously in anyone, yet so complex that it has puzzled even the greatest thinkers. It is a question so instinctive that it has intrigued human imagination since the dawn of civilization, yet so ambiguous that it leaves us in doubt as to whether it is a matter for religion, philosophy, or science. But it is indeed a question that underpins the whole of human knowledge. *How did the universe begin?*

Throughout history, humanity has never been so close to understanding the mechanisms that gave rise to the universe. While there are still plenty of unresolved mysteries, a coherent overall vision is finally emerging. This is the result of a subtle piece of detective work in which clues have been collected from every field of science—from the microscopic world of elementary particles to the sphere of astronomical observation—and then brought together to build up a picture of the early phases of the universe.

© The Author(s), under exclusive license to Springer Nature
Switzerland AG 2024
G. F. Giudice, *Before the Big Bang*, Copernicus Books,
https://doi.org/10.1007/978-3-031-69933-7_1

The science that studies the origin and evolution of the universe is called cosmology, the study of the *cosmos*, which in turn is Greek for order, harmony, and beauty. The word 'cosmology' actually has the same root as the word 'cosmetics,' which refers to the art of beautifying the body. Order and beauty provide the very reason why cosmology is able to make sense of the origin of the universe. The existence of a logical ordering of nature, a set of rules that can be expressed in rigorous mathematical language, is what guarantees the intelligibility of the universe.

Cosmological exploration has only achieved maturity in relatively recent times, thanks to some extraordinary progress in science. On the one hand, the discoveries of theoretical physics have revolutionized the way we understand space, time, matter, and forces, merging these concepts into a pattern so tightly knit that it is able to restrict the logically possible structures of objective reality. On the other hand, the development of new technologies has brought unprecedented progress in astronomical observation, with new experiments providing many unexpected and often sensational results.

The story I want to tell here is a journey back in time, to the very limits of human knowledge, and its destination is the origin of the universe. This journey is one of the most exciting adventures ever undertaken by humanity. Along the way, we shall be able to experience the magic of science and the pleasure of discovery, a pleasure that goes well beyond the bounds of science. Indeed, the question of the beginnings of the cosmos strikes deep chords in our sensibility and it is so vast as to encompass the entire sphere of human values.

The purpose of my story is to explain how the Big Bang came about. What exactly is it? And what caused this extraordinary event which, 13.8 billion years ago, gave rise to the particular blend of space, time, and matter that we call the universe?

In the course of our journey, we will discover the true nature of the Big Bang, reconstructed on the basis of astronomical data and logical deductions. We will thereby debunk the myth that the Big Bang was a colossal explosion that occurred at a tiny point of space. This cartoonish depiction of the Big Bang might have been fine at a time when young people wore long hair and flared trousers, but it is surprising that it remains intact in the collective imagination even now. For science has progressed since then, and today it tells a whole different story.

"A story should have a beginning, a middle, and an end," said the French film director Jean-Luc Godard, "but not necessarily in that order." In keeping with the canons of the *Nouvelle Vague*, my story will not scrupulously respect the temporal sequence of cosmic evolution, describing it phase by phase from birth to the present day. Instead, I will aim straight for the Big Bang, following the sequence of ideas that led humanity to this overwhelming discovery. So let us begin our story by looking at the way physics describes the reality of the cosmos and show how it has been able to unveil the extraordinary event that was so long hidden in the folds of time.

2

The Shape of the Cosmos

Others have seen what is and asked why.
I have seen what could be and asked why not.
Pablo Picasso

Master Mokurai was sitting in the courtyard of a Zen temple when he heard the gong announcing the arrival of his young student Toyo. After bowing three times, Toyo settled down on the cushion next to his master and remained there in perfect silence. Without taking his gaze from the horizon, Mokurai questioned the student with the following *koan*: "You know the sound of two hands clapping, but what is the sound of one hand clapping?" The boy retreated to his room, striving to listen to the silence with his eyes shut. He heard the music of the geishas, the sound of water droplets falling, the rustling of the wind, and the call of an owl. Left perplexed by these noises and his inability to find an answer to the *koan*, he withdrew into a deep meditation at a secluded hermitage. Much later, he came back to Mokurai with the answer: "I heard it, but it makes no sound."

There are questions which may get us thinking but seem to be outside the realm of science. However, the boundary between pure metaphysics and what can be addressed by the scientific method is in constant evolution as our understanding improves. Only a century ago, the question of the 'shape of the cosmos' was nothing more than a rather curious *koan*.

According to Newton, space and time serve as a silent theatre for the spectacle of physical reality. The real actors are matter and forces, for these determine the natural phenomena. Space and time are merely a stage upon which physical processes play out, but they are themselves strangers to action,

G. F. Giudice, *Before the Big Bang*, Copernicus Books, https://doi.org/10.1007/978-3-031-69933-7_2

remaining absolute and immutable. In the *Principia*, his essay of 1687 that would establish the foundations of physics for centuries, Newton declared: "Absolute, true, and mathematical time, of itself, and from its own nature, flows equably without regard to anything external. […] Absolute space, in its own nature, without regard to anything external, remains always similar and immovable."

This inflexible conception of space and time, although perfectly consonant with our everyday perception of the world around us, would be overturned in 1915 by the theory of relativity. According to Einstein, there is nothing absolute in either space or time. They are instead rather plastic entities. Einstein brought space and time to life, transforming them from a rigid backdrop into vibrant leading actors. Space and time react to physical phenomena. They are deformed and curved in the presence of matter or energy, and are in fact two facets of a single concept: *space–time*. In the relativistic view of things, space–time is a flexible entity shaped by the phenomena that occur within it, and yet at the same time capable of modifying those same phenomena. The symbiosis between the deformation of space–time and the motion of bodies is what we perceive as the force of gravity. According to Einstein's relativity, gravitation is just an illusion. It is the effect felt by a body moving freely in a curved space–time, deformed by the presence of matter or energy.

To produce a more tangible image, one can visualize the distortion of space–time due to matter as a kind of elastic sheet that stretches under the steps of an acrobat walking across it. Any object immersed in physical reality will feel the effect of these deformations, sliding down the slopes of curved space–time. As summed up so succinctly by physicist John Wheeler: "Space tells matter how to move; matter tells space how to curve." So this was goodbye to Newton's absolute and immutable space and time and welcome to the world of relativity, where physical events themselves can shape space and time.

Spatial distances and temporal durations depend on the position and speed of whoever measures them. For example, incredible though it may seem, according to Einstein's relativity, the clock on the bell tower of a church beats time faster than the watch on the wrist of a passer-by down by the church door. This is not because one of the two clocks is malfunctioning, but because time really does flow more slowly in places where the Earth's gravity is stronger. Needless to say, the effect is so weak on Earth that it is hard to detect: the wristwatch of a passer-by will lose about a nanosecond a day compared with the time recorded by the clock on the bell tower. Yet the effect is real enough and has actually been measured using very precise atomic clocks. Indeed, for the GPS to work correctly, this relativistic effect must be taken into account when synchronizing the flow of time on a satellite with that on Earth.

In his 1931 painting *The Persistence of Memory*, the eccentric Catalan artist Salvador Dalí, who often sought inspiration in new scientific ideas, painted clocks which seem to melt and drip down to places where relativistic time flows more slowly. In addition to influencing artists like Dalí, the idea of a dynamic space–time, shaped by its content of matter and energy, offers us the exciting prospect of calculating, from astronomical measurements, the 'shape of the cosmos' or, to be more precise, the geometry of space–time. But to understand what this means, it will be worth taking a little time to spell out what is meant by geometry.

A Strange New World

Dante Alighieri placed Euclid among the great spirits that were nevertheless condemned to an eternity in limbo. He couldn't have done more for Euclid, given that the illustrious Alexandrian mathematician was born more than three hundred years before the Christian era. Raphael depicted him in the Vatican fresco known as the *School of Athens*, where he features among the greats of antiquity, stooping down to draw circles with a compass. These artists had every reason to celebrate him because Euclid made a truly enormous contribution to human intellectual progress.

Euclid understood that, starting from five postulates—that is, five manifest truths that must be taken for granted—all the properties of geometric shapes in space could be deduced using only mathematical logic. This was a genuinely monumental result because it showed how all the relationships between points, lines, and distances followed from just five simple assertions. If these hypotheses can indeed be taken for granted, and they really do seem like self-evident truths, it turns out that geometric space is absolutely unique and could not have had any other form.

Among Euclid's postulates, one in particular attracted the attention of mathematicians over the centuries. In its modern version, Euclid's fifth postulate states that, in a plane, given a straight line and a point outside it, there is one and only one parallel line that passes through that point. Since antiquity, various mathematicians tried to prove that the fifth postulate was in fact deducible from the other four and therefore superfluous. But all these attempts failed.

At the beginning of the nineteenth century, the famous mathematician Carl Friedrich Gauss realized that, by giving up the fifth postulate, it was possible to construct geometries that differed from Euclid's, and yet were perfectly consistent from a logical point of view. Gauss never published these results, perhaps because the geometric space he had thus constructed in his mind seemed just too strange.

Almost twenty years later, the Russian Nikolai Ivanovich Lobachevsky and the Hungarian János Bolyai, working independently, rediscovered the existence of alternative geometries to the one Euclid had proposed. In a letter to his father, also a mathematician, Bolyai wrote: "Out of nothing I have created a strange new world." He clearly sensed the immense scope of the idea that Euclidean geometry might not be the only logical possibility for physical space. Unfortunately, his enthusiasm was damped by Gauss who, after reading the results of the young Hungarian mathematician, wrote to János' father: "To praise it would be to praise myself; the entire contents of the work, the path that your son has taken, and the results to which it leads, are almost perfectly in agreement with my own meditations, some going back thirty to thirty-five years."

In 1854, the German mathematician Bernhard Riemann gave a complete formulation of non-Euclidean geometries, setting out the mathematical language that is still in use today. Very few would have imagined then that these abstruse geometries could have anything to do with the physical space in which we actually live.

Simplifying Riemann's classification to some extent, we can consider three possible types of geometry. In the *spherical geometries*, there are no parallel straight lines. If we define a straight line as the shortest path connecting two distinct points, then in spherical geometries, two straight lines must intersect sooner or later. Spaces of this kind—which in a manner of speaking curl up on themselves—are characterized by the property that straight lines tend to converge. Conventionally, these spaces are said to have *positive curvature*.

It is not easy to visualize a three-dimensional curved space, but the task becomes easier when one dimension is ignored and we restrict ourselves to a two-dimensional curved surface. For example, consider the case of the Earth's surface and suppose it constitutes the whole of the space accessible to physical reality, so that nothing else exists outside of it. This is actually quite a realistic example, since humans usually only move in the north–south and east–west directions, but rarely dive down into the bowels of the Earth or embark on interplanetary journeys. To a first approximation, we do indeed live on a two-dimensional spherical surface.

On a spherical surface, straight lines run along great circles, that is, the largest possible circles. Examples of great circles on the Earth are the equator, two opposite meridians, or any other circle of equal length. It is easy to see that great circles always define the shortest path between any two points on a spherical surface. For example, Rome and Boston lie almost at the same latitude of 42°, but a plane flying from Rome to Boston will not follow the forty-second parallel. Instead, it will fly further north, along a great circle, precisely because this is the shortest route, even though it appears as a curved line in the atlas (see Fig. 2.1a). Because any two great circles will always intersect somewhere, parallel straight lines cannot exist on a spherical surface.

One feature of spaces with spherical geometry is that the sum of the angles of any triangle is always greater than 180°, which is quite different to what happens in Euclidean geometry. Here's an example. Draw a triangle with one vertex at the North Pole, two sides along meridians that form a right angle, and the third side along the equator. Then all three sides follow great circles (see Fig. 2.1b). Moreover, each of the three internal angles in this triangle is 90° and their sum is 270°. So, let's face it: geometry on a spherical surface is really rather bizarre.

A second kind of non-Euclidean geometry is the *hyperbolic geometries*. In this case, given any straight line and a point not on it, it is even possible to draw an infinite number of parallel straight lines passing through the given point. These spaces—which in a manner of speaking tend to open up—are characterized by the property that the sum of the internal angles of any triangle is always less than 180°. In mathematics, the curvature of these spaces is said to be *negative*.

Finally, between these spaces with positive or negative curvature, we recover ordinary Euclidean geometry, where the sum of the angles of any

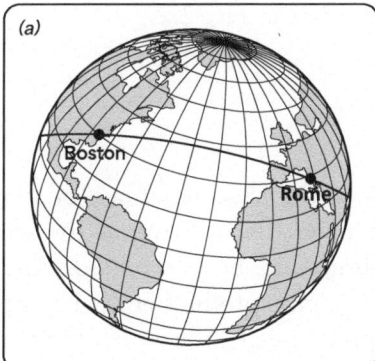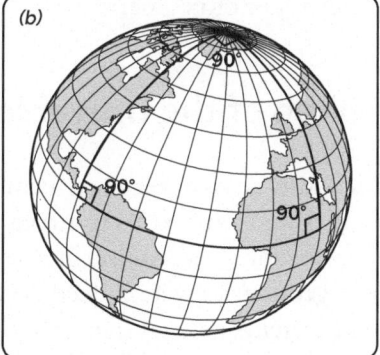

Fig. 2.1 **a** On a spherical surface, straight lines (taken as the shortest paths between pairs of points) correspond to what are known as great circles. On the Earth's surface, a straight line between Rome and Boston does not follow the forty-second parallel and therefore appears curved in an atlas. **b** On a spherical surface, we can draw triangles with the property that all their angles are right angles

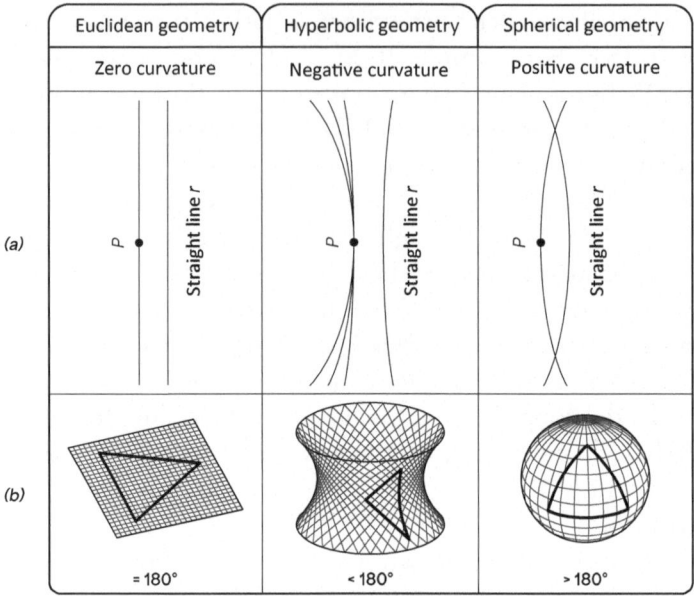

Fig. 2.2 **a** In Euclidean geometry, given a straight line *r* and a point *P* that is not on it, there is one and only one straight line passing through *P* and parallel to *r*. In hyperbolic geometry, there are infinitely many. In spherical geometry, there are none, because any two straight lines must necessarily intersect. **b** In Euclidean geometry, the sum of the internal angles of any triangle is 180°. In hyperbolic geometry, it is always less than 180°, and in spherical geometry, it is always more

triangle is always 180°, just as we learnt at school. In this case, physicists speak of *flat geometry*, in the sense that the curvature is zero. The properties of the three types of geometry are summarized in Fig. 2.2.

The Shape of the Universe

Non-Euclidean geometries may seem to be the obscure fantasy of some idle mathematician or an extravagant game played by looking at the world in distorting mirrors, like those in the fairground booths of old. And non-Euclidean geometries were indeed born as a pure product of the imagination. But then Einstein's general theory of relativity turned the tables, showing how space–time really does twist and curve, giving rise to forms that can only be described with the unfamiliar rules of non-Euclidean geometry. So, it turned out that non-Euclidean spaces are not at all an obscure fantasy, but a physical reality, despite everything our common-sense intuition might suggest. But why trust intuition anyway?

Our innate capacity for intuition is the fruit of Darwinian natural selection refined over thousands of years wandering in the savannah in search of food. Along the way, we have developed an extraordinary ability to estimate the parabolic trajectory of a stone thrown in our direction so that we may dodge it in time, and we have developed a peripheral vision that is particularly sensitive to the movement of any kind, so useful in avoiding an ambush. These abilities are all part of what we call intuition.

Newtonian physics largely encodes what our perception naturally suggests to us. Instead, relativity and quantum mechanics—the two cornerstones of modern physics—deal with phenomena that are completely alien to our everyday perception, governed by laws that defy common intuition. Knowing about relativity and quantum mechanics would not entail any evolutionary advantage. A caveman meditating on the motions of bodies near a black hole or capable of understanding quantum indeterminacy would soon be torn apart by the first wild beast to happen by, without having time to transmit his exceptional genes to future generations.

When we reflect on the origin of the universe, we will come up against physical phenomena that are foreign to everyday experience. As we delve deeper into the exploration of the Big Bang, we will increasingly have to deal with the strangest aspects of modern physics. So, it will be good to abandon that common-sense intuition which, although essential for the evolution of our species, can only lead us astray when we are dealing with the mechanisms of the Big Bang. In short, to understand the shape of the cosmos, it will be better to set aside our usual intuition and instead let ourselves be carried along by the mathematics of non-Euclidean spaces.

Non-Euclidean geometries allow for situations that are impossible in ordinary geometry. For example, it seems right to think that a space with finite volume must necessarily have a boundary. However, this is no longer true in a spherical geometry like that of the Earth's surface. Walking on the Earth, we never encounter boundaries, in the sense of places where space ends. At the most, we may return to our starting point. Although there are no boundaries, the space of a spherical surface is nevertheless finite.

Non-Euclidean geometries offer a quite different perspective when it comes to questions like: is physical space infinite? We are still unsure of the answer to this question, but at least we know we don't need to worry that space might end at some kind of Pillars of Hercules, beyond which there is nothing. Non-Euclidean geometries allow for space to be finite and yet have no end.

We are used to assessing the shape of an object by turning it over in our hands and observing it from all sides. In other words, we conceive its shape by the way it is immersed in space. But none of this works when we consider the shape of the cosmos, since the cosmos is itself the totality of space. The shape of the cosmos is not described by the way it is immersed in some even bigger space, but through its intrinsic geometry.

The examples of two-dimensional spaces shown in Fig. 2.2b are useful for forming a visual image of non-Euclidean geometries, but they should not be allowed to mislead us. In the figure, the two-dimensional surfaces are intended to describe the entire physical reality, while the three-dimensional space is an imaginary construction. To visualize curvature, we naturally immerse space in a hyperspace, that is, in a larger space. To visualize a curved line, we draw it on a sheet that serves as a two-dimensional hyperspace. To visualize a curved surface, we think of it as embedded in a three-dimensional hyperspace.

The revolutionary aspect of non-Euclidean geometries is the existence of curved spaces that are not embedded in anything else. A space can be curved without being the surface of anything. The curvature of physical space is intrinsic to the space itself and does not require any hyperspace. Physical space does not curve in an external space, but within itself.

Ultimately, understanding the shape of the cosmos means deducing the geometry of the whole of space–time. This geometry not only describes the curvature of space, but also tells us about the past and future of the universe. As soon as he had formulated the general theory of relativity, Einstein immediately understood that it provided him with a unique opportunity. The shape of the cosmos was no longer some kind of mystical *koan*, but a question that came within the ambit of scientific investigation. Einstein realized that he could play at being the creator of the universe, and he immediately set to work.

The central equation of general relativity relates matter and energy to the geometry of space–time. In other words, knowing how matter and energy are distributed, this equation tells us everything we need to know about the curvature of space–time, and that in turn describes all the effects of the gravitational force. In honor of its author, this equation is known in physics as Einstein's equation.

Einstein's equation is perhaps the most beautiful equation ever written in physics, because it expresses the deep connection between matter/energy and geometry/gravitation. It is so beautiful that it has been painted on the outer wall of the Boerhaave Museum (see Fig. 2.3). When physicists talk about Einstein's equation, they mean this one, and not the more popular but less fundamental equation $E = mc^2$.

$$R_{\mu\nu} - \frac{1}{2}Rg_{\mu\nu} + \Lambda g_{\mu\nu} = \frac{8\pi G}{c^4}T_{\mu\nu}$$

Fig. 2.3 Einstein's equation, as it appears on the outer wall of the Boerhaave Science Museum in Leiden

If only we knew how matter and energy were distributed in the universe, Einstein's equation would provide us with a unique answer to the question of its geometric shape. Therefore, to set up as creator of the universe, Einstein had to start from an assumption about the distribution of matter, and he chose the simplest one: he supposed that matter was spread throughout the cosmos in a continuous and uniform way.

The Universe Shaped Like a Spherical Cow

The idea that matter might be distributed in the universe in a continuous and uniform manner may seem absurd. Perhaps Einstein was too engrossed in his equations to bother looking at the night sky dotted with stars and realize that the universe is anything but uniform? Yet, Einstein's hypothesis was in fact perfectly sensible.

The art of doing theoretical physics often amounts to identifying hypotheses that simplify the problem enough to solve it, but not so much as to lose its essential properties. In other words, the idea is to eliminate everything that obscures the main features and get down to what really matters.

Physicists call this trick of the trade a *spherical cow approximation*. Seen from a distance through a low-resolution camera, a cow will look roughly like a sphere, provided one cannot make out the horns, the number of legs, or the length of the tail. Making a spherical cow approximation means capturing the essential properties of a problem, while disregarding irrelevant details.

This really is an art, because it is the result of intuition in being able to distinguish what is physical reality from what obscures its content. Inventing a spherical cow means painting an abstract picture of the universe. It is no coincidence that artists have grappled with the same problem, as so lucidly expressed by the American painter Georgia O'Keeffe in an interview in 1922: "Nothing is less real than realism. Details are confusing. It is only by selection, by elimination, by emphasis, that we get at the real meaning of things."

It was because the details had confused Aristotle that he thought a moving body would stop of its own accord. Galileo, on the other hand, formulated the correct laws of motion when he had the intuition that he should disregard the effects of friction. Physicists construct approximate models of reality which allow them to understand the meaning of a phenomenon more effectively than what could be achieved with a meticulous description. In a different context—but perhaps not so different after all—Pablo Picasso summed up the idea of the spherical cow approximation: "Art is a lie that makes us realize the truth."

The universe is anything but uniform when observed at terrestrial distances, and even at distances typical of the Solar System, galaxies, and clusters of galaxies. But astronomical images on larger scales, beyond about three hundred million light-years, do indeed reveal a general uniformity in the distribution of matter in the universe, without there being any particular structure to it.

The assumption that matter is spread in a continuous and uniform way is an excellent spherical cow approximation, if we are only interested in the overall shape of the universe and not the individual structures within it. This hypothesis corresponds to a blurred image of the cosmos, as if the entire universe were made up of a uniform gas of galaxies that could not be distinguished individually due to the poor resolution.

This blurred image of the cosmos is comparable to our way of perceiving matter. A block of marble appears to us as a uniform continuum of matter, even though in reality, when observed at distances of a billionth of a meter, it turns out to be a complex arrangement of atoms immersed in empty space. As long as we limit ourselves to studying the global properties of the block of marble rather than its microscopic structure, the description in terms of a continuum of matter is not only a valid approximation but is even the most appropriate way of dealing with it. Naturally, reality would look very different to tiny imaginary beings who live in the depths of matter, for they would see only empty space around them, dotted about with the occasional isolated atom.

When we look out upon the universe, we are like those tiny beings. We see stars and planets immersed in a vast and largely empty cosmic space. It is only through imagination that we can build up a picture of the universe as a whole and thus visualize its uniform structure. The purpose of cosmology is not to study individual celestial bodies, but the global properties of the universe, so the spherical cow approximation that takes the cosmos to be a uniform substance is an excellent starting point.

Einstein did not only assume the uniformity of matter in space. He added another hypothesis, namely that the global structure of the universe does not change over time. In those days, this seemed like a self-evident truth. What sense would there be in an evolving universe, with a beginning or an end?

Einstein was disappointed to find that, starting from the hypothesis of perfect uniformity in space and time, his equation did not admit any solution at all. His calculations showed that there could be no static universe in which matter is distributed continuously and uniformly.

Upon reflection, this result is not surprising at all. Indeed, Newton had already grappled with the same issue more than two hundred years earlier when he asked: since gravity is an attractive force, what prevents the universe from collapsing in on itself?

Newton believed he had found the answer in the infinite vastness of the universe. In correspondence with the theologian Richard Bentley, he argued that, if all the stars in the universe were distributed uniformly in a finite space, the force of gravity would make them collapse towards the center. But if space were infinite, there would be no center. Therefore, an infinite and uniform universe would remain static, because there wouldn't be a point towards which it could collapse.

Although ingenious, Newton's answer was wrong and shows how even a genius can be mistaken when dealing with infinities. Alas, an infinite universe uniformly populated with matter will in fact collapse under the effect of gravity. Although there is no center, the collapse occurs uniformly and any observer will see all matter converging towards them, regardless of where they are located. The collapse occurs both in Newtonian gravity and in general relativity and is the reason why Einstein did not find any static solutions to his equation.

Einstein did not at this point abandon his project of giving shape to the universe but chose to follow a different path. In doing so, he actually made a serious blunder. But, at the same time, he also made an extraordinary discovery. His mistake lay in not considering the possibility that the universe could change. By neglecting this possibility, he just missed the discovery of the Big Bang, which was sitting unequivocally inside his equation. Presumably, he considered the immutability of the global structure of the universe to be some kind of incontrovertible truth. The idea of a universe that swells up through a prodigious expansion of space, or one that deflates and crumples up miserably, must have seemed to him too far from reality. It is somewhat paradoxical that the physicist who undermined the rigidity of space and time, transforming them into malleable entities, suddenly stopped short of conceiving a universe in which space evolves dynamically. But the progress of scientific ideas does not always follow a logical path.

This great blunder was accompanied by a brilliant intuition: in his search for a solution that could make the universe static, Einstein introduced a concept that we believe today to be the key to understanding what happened *before* the Big Bang.

Einstein's Static Universe

At the beginning of 1917, when almost a million dead had been counted among the French and Germans on the Western Front, when chemical weapons such as chlorine, phosgene, and mustard gas were making their appearance on the battlefield, while the Dolomites witnessed in silence the steady loss of young Italian and Austrian lives, and when the United States was contemplating entering the war, Einstein was engrossed in his attempt to make the universe static. In February, he wrote to his friend and colleague Paul Ehrenfest that his efforts exposed him to the risk "of being confined in a madhouse." But instead of ending up in an asylum, he stumbled upon a solution that seemed fitting to him.

Einstein realized that a new term could be included in his eponymous equation, one that no one had previously considered. Unlike the term that described the matter and energy distribution, this new term had no clear physical interpretation. In fact, it remained entirely enigmatic because its effect was to produce a repulsive gravitational force, that is, the exact opposite of all known gravitational phenomena. The existence of a repulsive form of gravity, one might say anti-gravity, was an absolute novelty, something that could not even be contemplated in Newtonian theory. Despite there being no known anti-gravity phenomenon in nature, Einstein's equation was perfectly compatible with this new term, which became known as the *cosmological constant*.

Einstein used the unusual property of the cosmological constant to create a static universe, carefully balancing the gravitational attraction of matter with the gravitational repulsion it could provide. Just as in a tug of war between two equally strong teams, Einstein's universe remained in balance due to the equilibrium between these two opposing forces.

It had been well known since Newton's time that the gravitational force between two material bodies decreases as the square of the distance. In contrast, the repulsive force due to the cosmological constant grows linearly with the distance. This means that, as the distance doubles, the attractive force becomes four times weaker, when at the same time the repulsive force becomes twice as strong. For this reason, the effect of the cosmological constant is only important at great distances, while remaining negligible in our vicinity. This is the secret that allowed Einstein to create such a universe, without contradicting centuries-old astronomical knowledge that had never encountered anti-gravity. As long as we consider the distances relevant to typical astronomical observations, the effect of such a cosmological constant would be absolutely invisible. However, when we consider the universe as a whole, it would become the decisive factor in maintaining the cosmos in static equilibrium.

Visualizing four-dimensional space–time requires extraordinary imaginative skills. Fortunately, the hypothesis of a uniform cosmic matter distribution comes to our aid, simplifying our task. Since Einstein's equation relates matter and geometry, the uniform matter assumption implies that the geometry of space is also uniform. This means that, moving along any of the three spatial dimensions, we will always encounter the same geometry. We can therefore characterize space with a single dimension, which indicates the scale of distances between physical points, and suppress the other two. In the following, I will often use this trick, depicting space–time as a two-dimensional surface in which one direction indicates time and the other space. The reader will then be left with the imaginative effort—no easy matter—of recreating in their mind the two missing spatial dimensions.

The geometry of Einstein's universe is shown for a two-dimensional space–time in Fig. 2.4. It is a cylindrical surface, in which the vertical axis corresponds to time and the circles obtained by cutting the cylindrical surface with horizontal planes correspond to space. It is worth remembering that only the surface of the cylinder describes the physical space–time. The rest is there merely to visualize the curvature and is not part of reality. Figure 2.4 can be thought of as a stack of circles, which represent snapshots of space taken at successive instants of time as we move up the cylinder axis. The point is that all these snapshots are identical because the universe is static. Ultimately, Einstein's universe is a curved space of spherical type, whose geometry does not change over time. The radius of curvature of Einstein's spherical universe can be deduced from the observed density of matter in the universe, yielding a value of a few tens of billions of light-years.

Einstein was convinced that he had obtained the only possible universe consistent with the rigorous dictates of general relativity. Yet, from today's perspective, Einstein's universe appears untenable.

One of the main defects of Einstein's universe is the instability of the balance between matter and the cosmological constant, due to the different

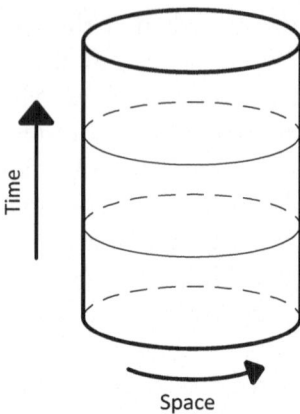

Space

Fig. 2.4 The shape of Einstein's static universe. The vertical direction indicates time while the direction around the cylindrical surface indicates space. Two of the three spatial dimensions have been suppressed. Horizontal sections of the cylindrical surface represent snapshots of space at different times. All the images are identical because this universe is static

dependence on distance of the two contrasting gravitational effects. Indeed, if the radius of the universe were only slightly greater than the equilibrium value, the cosmological constant would prevail and the universe would expand rapidly. Conversely, if the radius were only slightly smaller than the equilibrium value, the attractive effect of matter would win out and the universe would implode relentlessly. In other words, the balance that keeps Einstein's universe static is unstable and any small inhomogeneity in the distribution of matter would cause it to expand or collapse.

It's as if the tug of war were being played out on a high plateau and both teams had a precipice immediately behind them. At the slightest movement of the rope one way or the other, one team would fall into the abyss and the balance would be broken. In a similar way, Einstein's universe may be static, but this is only achieved by imposing an artificial and unrealistic condition.

Einstein's static universe was therefore a failure. Yet, the study that led Einstein to the creation of his abortive universe would be of great importance for our understanding of the Big Bang. Today we know that nature has made great use of the cosmological constant in its construction of the universe, even if it certainly did not use it to prop up the cosmos in the shaky state of equilibrium imagined by Einstein.

Happily, the idea of the cosmological constant survived the early demise of the static universe. In physics, ideas are often more vigorous than their inventors, and end up acquiring a quite different role from the one they were originally assigned. The cosmological constant is a perfect example.

De Sitter's Ineffable Space

Willem de Sitter was a reserved but determined man. In 1897, when he finished his studies in astronomy at the University of Groningen in the Netherlands, he became an assistant at the Royal Observatory of the Cape of Good Hope in South Africa. Arriving in Cape Town, he met Eleonora Suermondt, a young Dutch governess with a troubled family history: her father, a coffee grower in the colonies in Indonesia, had died when she was only eleven, her mother had been locked up in an asylum in Java, and her maternal grandfather had been a pirate in the South Seas, operating under the pseudonym of Monseigneur de Mérode.

Between Willem and Eleonora, it was love at first sight. He invited her on a trip to Table Mountain, the characteristic flat-topped mountain that overlooks Cape Town. Once they reached the top, he asked her to marry him. Eleonora, who remained in adoration of her husband for her entire existence, wrote in her old age that her life began that day on Table Mountain, where Willem had told her that he was "completely full of love, just as cut glass is full of light." For an astronomer used to work with lenses and telescopes, these words must have been the pinnacle of romanticism.

The wedding was arranged in something of a hurry. It is not known whether Willem's parents approved of the choice, but his father asked that a postscript be added to the marriage contract stipulating a regime with a separation of property.

De Sitter later became a respected astronomer at the University of Leiden. An uncle of Eleonora's commented that we should never despair when a child chooses a profession that we do not approve of, if indeed the niece of a pirate can marry a famous professor of astronomy.

In 1917, shortly after the publication of Einstein's static universe, de Sitter proposed an alternative model of the cosmos. In all modesty, he called it solution B, thereby distinguishing it from solution A, which had been conceived by the undisputed authority in the world of physics, who was also a dear friend of his, seven years his younger.

De Sitter started out from an approximation of the universe that differed from Einstein's. Since stellar matter is relatively rare in the universe, de Sitter's spherical cow approximation consisted in describing the universe as if it were empty, that is, completely devoid of matter. Like Einstein, de Sitter took into account the cosmological constant and required the universe to be static. Under these assumptions, he solved Einstein's equation and obtained solution B, which is known today as *de Sitter space*.

De Sitter's solution did not please Einstein at all. For his part, Einstein was firmly convinced that a single possible universe should emerge as an inevitable consequence of the general theory of relativity. The equations should admit one and only one solution, the one he had discovered himself. Einstein did everything he could to find flaws in de Sitter's calculations, but without success. He nevertheless told his friend and colleague that the new solution "does not correspond to any physical possibility." The debate about the two possible universes, Einstein's and de Sitter's, went on for a decade and drew in physicists and astronomers from all over the world.

To visualize de Sitter's universe, we may consider a single spatial dimension and suppress the other two, as we did before. Then, de Sitter's universe is represented by a two-dimensional space–time surface in the shape of a hyperboloid (see Fig. 2.5). If we interpret the vertical axis as representing time and intersect the hyperboloid with horizontal planes, we find that, as time goes by, de Sitter's space first contracts to a minimum radius and then expands inexorably. This result is somewhat puzzling, because it contradicts the initial assumption that the universe is static. So, what is happening? Is de Sitter's universe static or is it expanding?

The solution to the enigma lies in the fact that, since de Sitter's universe is devoid of matter, there is no preferred way to separate space from time. The space–time in Fig. 2.5 has an absolute meaning, independent of the position and motion of the observer, but what we call space and what we call time are the result of an arbitrary choice that depends on our particular point of view.

De Sitter's universe looks very different depending on the separate definitions that are made for space and time. In particular, there are three different

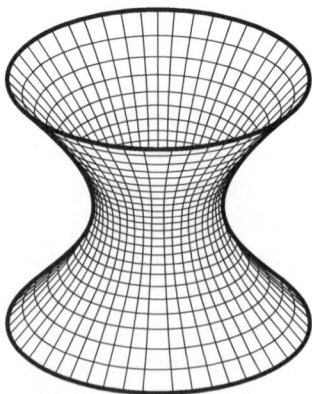

Fig. 2.5 The shape of de Sitter's universe. The two-dimensional surface of the hyperboloid describes space–time with a single spatial dimension, while the other two space dimensions are suppressed

ways of taking spatial slices of de Sitter's universe, all perfectly logical, and which correspond to expanding spaces with hyperbolic, or spherical, or even Euclidean geometry. In his original article, de Sitter had chosen yet a different way of slicing space–time, wherein the shape of space remained the same at every instant of time, making the universe appear static. In short, de Sitter's space is something of a chameleon! It takes on different forms depending on how we look at it. The reason for this is the high degree of symmetry intrinsic to de Sitter's empty space.

It may seem surprising that in 1917 neither de Sitter nor Einstein were aware of the various quite legitimate interpretations that could be made of the space–time described by the hyperboloid in Fig. 2.5. General relativity involves rather complicated mathematics and, when it was first proposed, some of its aspects were not necessarily obvious, even to those who had invented the theory.

The ambiguity about how to separate space from time is resolved as soon as even a minimal dusting of matter is added to de Sitter's empty space: the most appropriate choice is the one that guarantees the uniformity of matter at every moment in time. This choice is the most convenient for cosmological studies aimed at describing our actual universe, and it unequivocally fixes the definitions of space and time in the de Sitter world, even though it barely makes it less enigmatic.

From a cosmological point of view, the de Sitter universe is anything but static, despite its author's attempts to make it so. Instead, it is characterized by a dizzying expansion. And although space evolves, time has neither beginning nor end. Moreover, the rate of expansion of space is constant in time and a hypothetical inhabitant of de Sitter's world capable of observing only a limited region of space will always see an identical universe. Such an observer will have no way of distinguishing the past from the future. For a de Sitterian, although space expands, time has no meaning.

All this may seem like an abstract mathematical fantasy that describes an absurd universe. Yet the idea, born and then abandoned at the beginning of the twentieth century, found a new life about sixty years later and, as we shall see, plays a crucial role in revealing the mysteries of the Big Bang. Not for the first time in the history of science, an idea has gained the upper hand over its originator.

3

The Pioneers of the Big Bang

Two roads diverged in a wood, and I–
I took the one less travelled by,
And that has made all the difference.
Robert Frost

In 1848, shortly before his premature and mysterious death, the American writer Edgar Allan Poe published *Eureka*. It was not a horror story or a detective novel—the themes that made Poe famous—but a "prose poem," to use Poe's own words, an essay describing his personal conception of the material and spiritual universe.

Poe hated the idea of an eternal and mechanistic universe. He was convinced that rational thought and the deductive scientific method would never be enough to understand the deep reality of the physical world. To penetrate the secrets of nature, a new, more spiritual tool would have to be found, and the American writer identified this to be the human faculty of intuition.

Eureka is an ambitious but nonsensical tale, in which dreamlike visions of the universe are described as if they had a scientific basis. Perhaps precisely because of its obscurity, the text is considered by some modern critics as a precursor of relativistic concepts and cosmological ideas. In the story, Poe imagines that the universe sprang from an instantaneous flash which, out of the explosion of a primordial particle, generated all matter: "From the one Particle, as a center, let us suppose to be irradiated spherically—in all directions—to immeasurable but still to definite distances in the previously vacant

G. F. Giudice, *Before the Big Bang*, Copernicus Books, https://doi.org/10.1007/978-3-031-69933-7_3

space—a certain inexpressibly great yet limited number of unimaginably yet not infinitely minute atoms."

Poe was convinced that *Eureka* was his most important work, for which he would be remembered forever by posterity. This would not be the case. In 1940, when the writer Arthur Quinn asked Einstein for his opinion, the latter compared *Eureka* to one of the many letters he received from cranks, almost on a daily basis, attributing its content to the "pathological personality" of its author.

Only someone completely deranged could propose a scientific theory in which the universe had a beginning. The idea was perfectly antithetical to the view of the physical world that had been built up through centuries of scientific progress. The Nobel laureate in chemistry Walther Nernst even argued that "to deny the infinite duration of time would be to betray the very foundations of science."

It would take two highly unconventional individuals, according to the standards of the academic world, to see what others refused to see. Two individuals who stopped to listen to what Einstein's equation was actually telling everyone, and who chose to follow a different path from the one explored by others. And this made all the difference.

The Universe's Unlucky Weatherman

While Einstein and de Sitter, the two titans of relativity, debated the pre-eminence of their respective universes, a mathematically inclined Soviet geophysicist and meteorologist, unknown in the West, made a truly fascinating discovery.

Alexander Alexandrovich Friedmann was born in St. Petersburg in 1888, grew up in Petrograd, and died in Leningrad, whence he spent most of his life in the same city. These were turbulent years in Russia. The son of a composer and dancer of Jewish descent and a pianist of Czech origin, he showed no musical talent, but a strong propensity for mathematics. His parents divorced when Alexander was nine years old, and the boy went to live with his father, who died just a decade later. At the outbreak of the First World War, Friedmann volunteered for the air force and was later appointed professor in the Department of Mathematics and Physics at the University of Perm, the city on the edge of the Ural Mountains that gave its name to the Permian geological era. In 1920, he returned to his hometown with a chair at the Geophysical Observatory, but he also took up positions in other institutes to supplement his meager salary.

He quickly became one of the city's most esteemed scholars, known for his eclectic originality and mathematical rigor. He was a sensitive and generous man, who loved to praise his colleagues publicly. Speaking of a study in fluid dynamics on the way vortices wrap around themselves, he attributed all the credit to a collaborator, adding that the work was completed while "I was scratching my right ear with my left hand."

In a university exam, a student had a panic attack and struggled to give coherent answers. "Well," said Friedmann, "give me your examination card." He wrote: "Passed," then added: "Have you now calmed yourself? Everything is all right, isn't it? Then take your seat and answer these questions." He continued the exam, whereupon the student answered all his questions with great lucidity and passed brilliantly.

In the troubled social climate of Russia, Friedmann took refuge in an enthusiastic study of an exciting novelty in mathematical physics that had recently reached a war-weary Petrograd: Einstein's general theory of relativity. "No, I am an ignoramus, I know nothing," he confessed to his friends, "I should sleep less and should not do anything outside science, because all this so-called life is a mere waste of time."

In 1922, Friedmann published a study in which he classified all the solutions of Einstein's equation under the assumption that matter is distributed continuously and uniformly throughout the universe, just as Einstein had done, but without assuming in addition that the universe must be static. This was the key point. Friedmann abandoned the idea that the universe had to be immutable, leaving it up to the equations of general relativity to decide the fate of the cosmos.

Friedmann showed that Einstein's equation predicts three classes of possible universes. The three different possible shapes of space–time are shown in Fig. 3.1. As before, only one space dimension is displayed, indicating the scale of distances, while the other two are suppressed.

It is important to stress once again that, in these figures, only the two-dimensional surfaces describe the physical space–time, while everything else in the drawing is fictitious. For simplicity, I will only consider the case where there is no cosmological constant, although Friedmann also studied the more general case.

The first set of solutions yields the class of *closed universes*, which are obtained when the matter distribution is denser than a certain critical value. The universe is closed in space, i.e., it always has a finite volume, and also in time, i.e., it has a limited duration. It is described by a curved space with spherical geometry that is born at the instant of the Big Bang, expands to

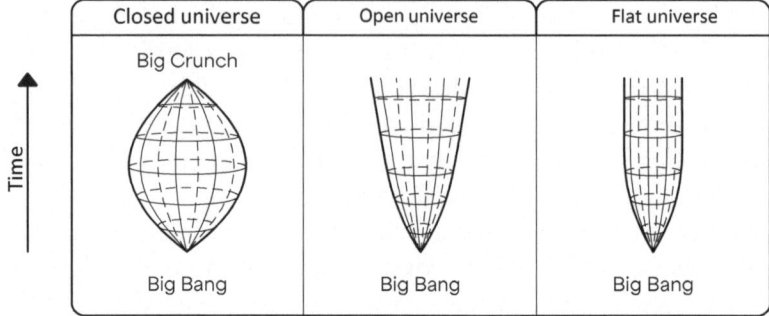

Fig. 3.1 The three kinds of Friedmann universe: closed, open, and flat. Time flows from bottom to top and cross-sections of the surfaces by horizontal planes indicate the distance scale between points of space

reach a maximum size, and then implodes to annihilate itself in a Big Crunch, a kind of Big Bang in reverse.

It is easy to understand the evolutionary behavior of a closed universe using a simple analogy. A stone thrown upwards moves away until it reaches a certain maximum height. At that point, gravitational attraction gets the better of it, whereupon the stone reverses its motion and falls back to Earth. In the same way, the radius of a closed universe grows after the Big Bang, but the expansion is slowed down by the gravitational attraction exerted by matter. At a certain moment, gravity gets the better of the expansion and the universe begins to fall in on itself, until it reduces to nothing in the Big Crunch.

The second set of solutions Friedmann found yields the class of *open universes*, which are obtained when the matter distribution is less dense than the critical value. The universe is open in space, i.e., its volume can become infinite, and also in time, i.e., its duration is unlimited. It is described by a curved space with hyperbolic geometry that is born at the instant of the Big Bang and expands forever. However, the rate of expansion decreases over time and converges on a limiting value.

For an open universe, we can use the same analogy as above, with one difference. If the stone is now thrown upwards with a speed greater than a certain critical value (in physics, this is called the escape velocity), gravity will slow it down, but will never be able to bring it to a halt. As long as we can neglect air resistance, the stone will continue its journey into space without ever turning back. In the same way, the expansion of an open universe slows down, without ever completely stopping.

The third case is the intermediate condition between closed and open universes. It is called a *flat universe*. This occurs when the density of matter is exactly equal to the critical value that corresponds to the minimum for a closed universe and the maximum for an open universe. The geometry of space is Euclidean, but the distances between physical points increase at a speed ever closer to zero as time goes by. The space of the flat universe can be visualized as a perfectly uniform elastic fabric, pulled equally in all directions by the expansion.

A common feature to all three classes of universes is that the solutions cannot be extended indefinitely far back in time. This means that there must necessarily be a beginning, an origin. There must be a Big Bang. This is a truly sensational result: it means that the Big Bang is a practically inevitable consequence of general relativity. Here we encounter one of the most incredible achievements of human knowledge. So incredible that, at the time, very few believed it.

For his part, Einstein did not take it well. Given that the very existence of the de Sitter solution had already irritated him, because it clashed with his conviction that a unique physical reality must emerge from fundamental principles, imagine his reaction to finding that there might actually be an infinite number of possible universes. He even went so far as to publish a paper on the spot, arguing that Friedmann's calculations were erroneous, because according to him they violated the basic principle of energy conservation.

However, Friedmann was absolutely convinced of the reliability of his results and wrote a letter to Einstein in which he very politely explained his objections and invited him to publish a correction, should he finally agree with Friedmann's arguments. He awaited a response somewhat anxiously, but heard nothing for almost six months. He could not have known that Einstein had not even seen the letter because he was traveling around Europe, and then went on to Japan and Palestine. Finally, in May 1923, Einstein came to Leiden University, where he met Yuri Krutkov, a friend and colleague of Friedmann's. Krutkov explained Friedmann's objections to him. At the end of the conversation, the inventor of relativity admitted his mistake and Krutkov wrote home: "Petrograd's honour is saved!" Einstein published the correction, although in the letter to the editor he wrote a sentence, cut out at the last moment, in which he had claimed that Friedmann's solutions, albeit mathematically correct, were irrelevant to the question of physical reality.

Friedmann's results had nothing of the resounding impact one might expect. They were largely forgotten and rediscovered only much later. The Soviet scientist discontinued his studies in cosmology and went through a period of severe depression. He had started up a relationship with Natalia Yevgenyevna Malinina, a young geophysicist at his own institute. He was tortured with guilt toward his wife Ekaterina, a woman of high moral standing who had always been dedicated to him, and his passion for Natalia.

"On my trajectory, as a symbol of the extreme points of my oscillations, stood you and Ekaterina," he confessed to Natalia. At the height of despair, he wrote to her: "I can't commit suicide now; I don't have the courage." He nevertheless decided to divorce Ekaterina and marry Natalia, with whom he had a son he would never know.

In 1925, Friedmann was appointed director of the Geophysical Observatory and accepted to undertake a mission aboard a hot air balloon to study the upper layers of the atmosphere. On the morning of 18 July, he and the pilot set off in decidedly unfavorable weather conditions. Friedmann's written report had nothing of a dry list of data but contained pages of almost poetic description of his feelings as he entered thick clouds, and the sights and sounds of human activity at the Earth's surface were gradually blotted out. At 6000 m, the two began to use the oxygen respirator. Accidentally, Friedmann caused a leak that generated an explosion, but miraculously, the balloon did not catch fire. The scientist, with his usual generosity, insisted that his companion use the remaining oxygen, while the pilot, faithful to Soviet values, proposed the same for his director. In a state of semi-consciousness and at times passing out completely, they reached an altitude of 7400 m, beating the previous Soviet record of 6400 m, which dated back to 1910.

In his report, Friedmann describes how, during the descent, he began to worry about where they would land. Would they come down on the beloved soil of the Soviet Union, where they would be welcomed by friendly compatriots, or bourgeois Estonia where they might perhaps be imprisoned? The wind was favorable and pushed them into a remote corner of the province of Novgorod, where many peasant women fainted upon their arrival, while certain brave young men helped the 'extraterrestrials' who had fallen from the sky.

Having completed this extraordinary endeavor, Friedmann hurried to reach Natalia, who was spending her pregnancy in Crimea. "It gives me joy to think that thousands of miles away the beloved heart is beating, the gentle soul is living, the new life is growing," he wrote to her, "the life whose future is a mystery and which has no past." Although they were clearly written in anticipation of their child, these words seem almost to evoke his visionary conception of the universe. Back in Leningrad, he suddenly fell ill and died within a few days, probably from an incorrect diagnosis of typhoid fever. Natalia learned the news from a newspaper.

Friedmann departed this life at thirty-seven years old, only three years after the publication of his solutions proving the possibility of evolving universes. Despite the fact that he did not receive any immediate recognition in the West, he was well aware that his discovery was quite revolutionary. A passionate admirer of the *Divine Comedy*, he loved to describe his research activities in Dante's words: "The water I cut was never sailed before." Friedmann had indeed been navigating waters never traversed by any before him.

As a partial reward for his unfortunate life, Friedmann today enjoys considerable fame among physicists. His merit lies in having dared to look beyond the limitations of static universes. His failing, if any, lies in not having sought to develop a complete cosmological theory, or investigate the connection with astronomical observations. For this reason, it would be going too far to identify him as the father of the Big Bang.

The Priest and the Primeval Atom

If one were asked about the personality of a typical cosmologist, one could expect almost anything. Except that it might be a witty, jovial Catholic priest who loves wining and dining. Yet one of those who came up with the idea of the Big Bang was more like Father Brown from the famous series of detective novels by G.K. Chesterton than a stereotypical scientist.

Born in Charleroi, Belgium, Georges Lemaître studied at a Jesuit college before taking up engineering studies at university. In the summer of 1914, he had planned a cycling holiday in the Tyrol with his brother and a friend, when Germany invaded Belgium, even though it had declared itself neutral. Just five days after the invasion, Lemaître enlisted.

At the end of the war, he decided to take up studies in physics and mathematics at the University of Leuven and, at the same time, enter the seminary. In 1923, after graduation and ordination, he obtained a scholarship to Cambridge, where he studied under the supervision of Arthur Eddington. A relativist and astronomer with a charismatic personality, Eddington was famous for a trip he made to the island of Príncipe, in the middle of the Gulf of Guinea, during a solar eclipse. There he led the mission that brought the theory of relativity to the attention of the general public, by confirming its prediction that light would be deflected by a massive object.

After his studies in Cambridge (England), Lemaître went on to Cambridge (Massachusetts), where he entered Harvard University and then the prestigious MIT. Following this important international experience, in 1925, he returned to Belgium where he had obtained a position at the University of Leuven.

He had something of a reputation for his unconventional way of lecturing. Rarely did his lectures follow any predetermined pattern. He preferred to improvise, jumping from one topic to another, quoting texts from Gauss and Jacobi in the original Latin and then perhaps illustrating a problem he had encountered the day before. Rather than teaching a subject, he taught his students an approach to the world of research.

The students, often bewildered by his lectures, made him the target of good-natured satire in the *Revues*, the traditional end-of-year skits. As he was a gourmet whose favorite radio program was *La minute gastronomique*, he was teased for indulging the alleged cardinal sin of gluttony. He defended himself by saying: "Some things prohibited on Earth are allowed in heaven!" When accused of the cardinal sin of sloth, he replied that a quality of the great mathematicians is that they "always reflect a lot before not computing." These self-parodies reveal his cordial nature and it is easy to understand the affection that students and colleagues must have felt for him.

To make up for the shortcomings of his extravagant teaching methods, he was generous with exam grades—after all, the gates of heaven are open to everyone. Sometimes he completely forgot to set the exam. One day he returned particularly late from the Majestic—his favorite restaurant—and found a group of puzzled students waiting in front of his office. Realizing that he should have held the exam more than an hour earlier, he shouted out: "Everyone passes with a 13!."

In 1927, Lemaître independently rediscovered the solutions, already found by Friedmann, that describe expanding universes. Extrapolating the evolution back in time, he deduced that there must have been a "beginning of the world," as he called it. The term 'Big Bang' was yet to be coined. Lemaître went further than Friedmann, deriving an equation that gave the recession speeds of galaxies in an expanding universe. This equation would be confirmed a couple of years later by the astronomical observations of Edwin Hubble. Somewhat unfairly, this law is known today as Hubble's law.

Furthermore, Lemaître sought an explanation for the physical phenomenon that could have set the expansion in motion. He wanted to understand what exactly could have brought about the beginning of the world, the event we now call the Big Bang. Unfortunately for him, the explanation would only come fifty years later. It should also be said that Lemaître was not really equipped for the task, because he had no real expertise in quantum mechanics and nuclear physics, the tools he would have needed to understand how matter behaves under the high-density conditions expected in the vicinity of the Big Bang.

He hypothesized that the initial state of the universe was a "primeval atom," whose nucleus contained all the matter and all the energy existing in the universe today. Radioactivity would have produced the initial spark which, by splitting the primeval atom into ever smaller atoms, gave birth to matter as we now know it. According to Lemaître, during the existence of the primeval atom, the concepts of space and time were meaningless, since the entire universe was concentrated within a single quantum element. He suggested that the chain of radioactive decays of the primeval atom was what drove the expansion of space, catapulting matter at high speed in every direction.

Unfortunately, the theory of the primeval atom is not based on solid mathematical foundations. It is merely a description, as Lemaître himself was quick to admit. Indeed, from the point of view of quantum mechanics, the idea is nonsensical. Lemaître's scientific merit lies not in his primeval atom—even though he was rather attached to this story—but in having identified the expansion of the universe, the recession of the galaxies, and the Big Bang as real physical features of the cosmos.

The brilliant Belgian physicist should also be credited with the remarkable insight that quantum mechanics might play an essential role in the origin of the universe. On this point he was right. As we shall see later, quantum mechanics is indeed an indispensable ingredient in the cosmic recipe.

Naturally, Lemaître was eager to know what Einstein thought about all this, but the guru of relativity did not give him a more favorable reaction than the one he had reserved for Friedmann. Lemaître first met him at the fifth Solvay Conference, held in Brussels in October 1927. During a walk in Leopold Park, Einstein listened attentively to his young colleague's explanations about the expansion of the universe and the beginning of the world. When Lemaître was through, he stopped, looked him in the eye, and said, "Your calculations are correct, but your physics is abominable." The two met several times in the years to come and maintained a sincere friendship.

Lemaître always wore the cassock and it is curious to see a priest portrayed alongside the familiar faces of physics in the photos of the time. His penetrating gaze framed by his round glasses and the stiffness of his posture seemed to betray a certain embarrassment when he had to pose next to the relaxed figures of Einstein or Eddington, who were much more accustomed to the indiscreet lenses of photographers. In reality, Lemaître felt absolutely at home in the scientific community. He always took an active part in discussions with other physicists and never held back on a witty remark when it was appropriate.

He strictly separated his scientific activity from questions of faith and saw no incompatibility. At the end of one of his public lectures, a spectator asked him if the primeval atom should be identified with God. Lemaître responded with a broad smile, adding, "I have too much respect for God to make Him a scientific hypothesis."

As he grew older, he gradually moved away from cosmology and developed a growing interest in numerical methods that could be applied in astrophysics. In 1933, he acquired several Mercedes electromechanical calculating machines, which were cutting-edge at the time. At the Brussels Expo in 1958, he discovered the Burroughs E101, and persuaded the university to buy one, contributing personally to the expenses. The E101 was a large desk-size electronic digital computer, programmed by inserting needles into holes, while the results appeared as numbers indicated by tiny light bulbs. To see it today, it looks like it came out of an old low-budget science fiction film.

In 1966, following a heart attack, he was admitted to the Saint Pierre Clinic in Leuven, where he was diagnosed with leukemia. His former assistant and friend Odon Godart went to visit him in hospital, bringing him news of the discovery of the cosmic background radiation. As we shall see later, this provided definitive proof of the existence of the Big Bang. Although Lemaître had argued throughout his life that it would be proven by identifying the particles emitted when the primeval atom split to produce cosmic rays, he immediately understood the significance of this discovery and was of course highly satisfied. Two days later he fell into a coma from which he never awoke.

4

The Unfolding Universe

It's the best possible time to be alive, when
almost everything you thought you knew is wrong.
Tom Stoppard

The French called them the *Années folles* (the crazy years), the Americans the *Roaring Twenties*. Rapid economic development after the First World War brought prosperity to Europe and the Americas, despite ominous signs of future nationalisms and dictatorships. Jazz music was all the rage, radios were appearing in people's homes, and the first films were being shown in theaters. While Howard Carter was uncovering the sarcophagus of Tutankhamun and Charles Lindbergh was flying across the Atlantic, quantum mechanics was revealing unimaginable secrets hidden in the depths of matter. Yet, cosmology was languishing.

Until the end of the 1920s, there was a general skepticism toward the idea that the universe could evolve. The results of Friedmann and Lemaître were practically unknown and the belief that the universe was immutable seemed itself to be immutable. In physics, however, experimental data matter more than opinions.

As early as 1912, the American astronomer Vesto Slipher at the Lowell Observatory in Arizona had measured the speeds of the nebulas, the name then given to those celestial bodies which we now know to be clusters of stars and which we call galaxies. The speeds of these nebulas were surprising. Peaks of several million kilometers per hour were measured. Never before had celestial objects been found to be moving at such insane speeds, and the news made headlines in the newspapers of the time. A particularly curious fact was

G. F. Giudice, *Before the Big Bang*, Copernicus Books,
https://doi.org/10.1007/978-3-031-69933-7_4

that forty-one of the forty-five nebulas analyzed by Slipher were moving away from us. This was indicative of a general recession, pushing the nebulas away from the Solar System, rather than a random distribution of speeds with each nebula just going its own way. Some physicists began to suspect that this unexpected behavior of the nebulas might have something to do with the strange properties of de Sitter's universe. But to clarify the matter, further astronomical observations were needed. And this is where Hubble came into the picture.

Edwin Hubble was destined for success. Over six feet tall, good-looking, intelligent, and self-confident, he excelled in every activity he turned his hand to. A champion in athletics, basketball, and water polo, Hubble was also a successful amateur boxer and it is said that he was once challenged to a fight with the then heavyweight champion Jack Johnson.

Despite being born and raised in rural Midwest America, he loved to affect a British accent, smoke a pipe, and dress like an English gentleman, so in many photos of the time he brings to mind a kind of Sherlock Holmes in the guise of astronomer. His ambition, egocentrism, and craving for social ascent were almost as vast as the universe he wanted to explore. He quickly made a name for himself in the field of astronomy and, by the age of thirty-five, he had achieved the fame he was so assiduously pursuing by identifying individual stars in the spiral arms of the Andromeda nebula. This observation was the definitive proof that the nebulas were not clouds of cosmic dust as their original appellation suggests, but huge clusters of stars held together by gravity. It turns out that we live in one of these clusters, the Milky Way galaxy, which is similar to many others.

Today we know that the observable universe is populated by millions of millions of galaxies, each containing a number of stars ranging from hundreds of millions to hundreds of millions of millions. Hubble's discovery thus opened a new perspective on the universe, showing it to be enormously larger than previously thought.

Hubble carried out his observations on nebulas, now called galaxies, at the Mount Wilson Observatory in California. Slipher had measured the speeds of a large number of galaxies, but their distance also had to be measured to understand whether their motion was caused by the expansion of the universe. Now, deducing the speed of a celestial body is a relatively simple task for astronomers, because the frequency of light carries a clear indication of how fast a star is moving away from us or toward us (using what physicists call the Doppler effect). But deducing the distance is a quite different matter. Hubble used an ingenious method invented years earlier by the American astronomer Henrietta Swan Leavitt. The idea was to exploit the rather special properties of a class of pulsating stars called Cepheid variables. The distance of such stars from the Earth can be deduced from their observed brightness and period of pulsation.

Many of the observations regarding the relationship between the galactic distances and recession speeds were made by Hubble, together with an unusual collaborator: Milton Humason. He had left school with only a fifth-grade education. As he loved the mountains, he got himself hired as a mule driver to transport materials for building the observatory at the top of Mount Wilson. When the construction work was completed, Humason was taken on at the observatory as a caretaker and cleaner. Out of pure curiosity, Humason got interested in the telescopes, and before long became a meticulous and tireless astronomer, making an important contribution to Hubble's observations. In 1950, he received an honorary doctorate in astronomy from the University of Lund in Sweden, becoming—as far as I know—the first person in the world to go directly from a fifth-grade education to a Ph.D.

The aim of Hubble's measurements was to show that the recession speeds of the galaxies increase in proportion to their distance from us, a result known today as Hubble's law (or seldom, but perhaps more fairly, the Hubble–Lemaître law). The experimental confirmation of this law, announced in 1929, was the blast needed to shake the prejudice that the universe had to be static.

But why does Hubble's law tell us the universe is expanding? And what do galactic motions have to do with the Big Bang?

The Expansion of the Universe

Hubble's astronomical measurements showed that the further the galaxy, the faster it moves away from us, with a law of direct proportionality. To see what this means, it will be convenient to introduce some rather unusual physical units. Instead of measuring distances in miles and speeds in miles per hour, I will express distances in millions of light-years (which I denote using the letter D, for brevity) and speeds in millions of light-years per billion years (which I denote by D/T). This choice is useful only because Hubble's results can then be expressed in small numbers, but the actual physics would be no different if I used miles and miles per hour.

Roughly speaking, Hubble measured that a galaxy at a distance of 1 D (i.e., one million light-years) moves away from the Earth at a speed of 1 D/T (i.e., one million light-years per billion years). Moreover, he measured that a galaxy at a distance of 2 D moves away at a speed of 2 D/T, while a galaxy at a distance of 3 D moves away at a speed of 3 D/T, and so on. This, in layman's terms, is the essence of Hubble's law.

Now imagine reversing the motions of the galaxies for a time T (i.e., one billion years). A galaxy that is 1 D away from us today and hence moving away at a speed of 1 D/T, must have been very close to us a billion years ago. But exactly the same applies to any galaxy at a distance of 2 D with speed 2 D/T, and the same again for one at a distance of 3 D, and so on. In other words, Hubble's law tells us that, if we wind the film of the universe backward for a time T, all the galaxies that are now flying away from us at these various speeds must originally have been sitting at practically the same point. That moment corresponds to the Big Bang.

The expansion of the universe is one of those discoveries able to revolutionize human thought. It tells us that the shape and size of the cosmos are in perpetual transformation. In other words, all physical phenomena occur within a dynamical structure that is itself in continuous change. The universe is not an absolute, unchanging, incorruptible entity. Like a living being, the universe evolves, grows, ages, and changes in appearance. The discovery of the expansion of space has completely overturned our previous conceptions of the cosmos.

But just a moment. What does space actually expand into? It is worth making the point clear: this question springs from a misunderstanding. We are used to the fact that creating space for ourselves means taking it away from someone else. But things don't work that way for the universe. We should not think of the expansion of the universe as something filling a void. There is nothing outside of space. Space expands in the sense that the distances between different points dilate over time. What we have here is not an expansion *into* space, but an expansion *of* space.

An effective way to visualize the expansion of space is to think of a map on your computer screen, whose scale can be increased by moving your fingers apart on the touchpad. As the scale changes, the overall geography remains the same, but regions expand, while towns and villages move away from each other.

The recession measured by Hubble is not due to any motion of the galaxies in space, but to the expansion of the space in which the galaxies live. Although the concept is not at all intuitive (almost nothing in relativity is), it is essential to keep in mind the exact meaning of the expansion of space if we are to avoid making mistakes in the interpretation of the Big Bang.

But if all the surrounding galaxies are flying away from us in a general motion of recession, does this mean that Aristotle was right in believing that we are exactly at the center of the universe? Absolutely not. Wherever we are in the universe, the recession will look exactly the same. To see this, it is useful to give an example.

Imagine a group of friends living on a constantly expanding planet. At regular intervals, Ada takes selfies that show her friends steadily moving away, even though none of them are walking (see Fig. 4.1). Ada sees Eva moving away from her twice as fast as Cloe, precisely because Eva is twice as far away as Cloe. Hence, Hubble's law applies and the recession speed is proportional to the distance.

Even though she sees everyone moving away from her, Ada is not in such a privileged position as to consider herself at the center of the universe. If Cloe were to take the selfies, she too would consider herself to be stationary at the center, while all her friends were moving away at speeds proportional to their distances. Neither Ada, Cloe nor any of their friends are at the center of anything.

In homogeneous spaces without boundaries, any observer at any point in the universe will see all the galaxies moving away at speeds determined by Hubble's law. The expansion of the universe is an expansion without a center. The center of the expansion is, at the same time, everywhere and nowhere.

In the film *Annie Hall* by Woody Allen, the main character, as a child, falls into a deep depression when he reads in the newspaper that the universe is expanding. Taken by his mother to the inevitable New York psychoanalyst, the child explains: "Surely this means that Brooklyn is expanding, I'm expanding, you're expanding, we're all expanding." The psychoanalyst tries to reassure the child with trivial arguments. For my part, I would have tried to reassure him by explaining what straightforward physics has to say about it.

To begin with, the expansion of space is a cumulative effect that becomes important only at great distances. A distance equal to the height of a man

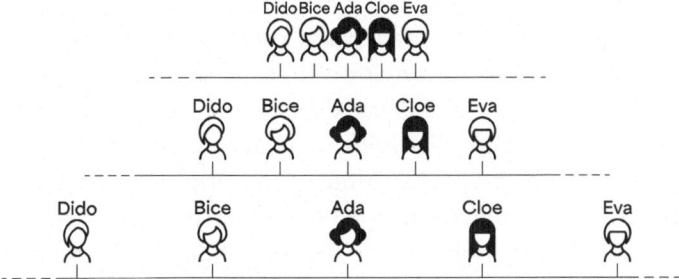

Fig. 4.1 Selfies taken at equal intervals of time by Ada, who lives on a constantly expanding planet. Ada sees her friends moving away at speeds proportional to their distances, as predicted by Hubble's law. Ada's observations are the same as those Cloe would make, or anyone else taking the selfies

will grow by about a nanometer in ten years. This is about the size of an organic molecule. And in any case, our body does not undergo even this tiny effect. The bonds that hold molecules together in solid bodies are much stronger than this weak gravitational effect, so the expansion of the universe does not change the internal structure of matter at all. In conclusion, neither Brooklyn nor its inhabitants are expanding due to cosmic evolution and there is no reason for anyone to get depressed. I am sure this explanation would have cheered up the young man in Woody Allen's film, a demonstration that physics has a far superior therapeutic effect to psychoanalysis.

Hubble's law might raise a further doubt. If the recession speed of galaxies increases with distance, at a certain point it will exceed the speed of light. So, isn't there then a contradiction with relativity, which forbids superluminal speeds?

Einstein's theory asserts that no physical information can be transmitted faster than the speed of light. The expansion of the universe is just a stretching of space without any exchange of information between different points. Therefore, a superluminal expansion of space does not violate any of the principles of relativity. Here is an example that may help to clarify the issue.

Imagine shining a laser beam on a distant screen, rotating the pointer at a constant rate, like a lighthouse by the sea (see Fig. 4.2a). The laser beam produces a small bright spot that moves uniformly across the screen. Moving the screen further away, the spot will move faster, with a speed directly proportional to the distance of the screen, just like Hubble's law (see Fig. 4.2b). Moving the screen further and further away, the speed of the spot will eventually exceed the speed of light. Does this superluminal speed disprove Einstein's relativity?

The answer is no, because the motion of the spot on the screen does not propagate physical information. The relationship of cause and effect follows only the laser beam, traveling at the speed of light, and not the bright spot. In fact, the propagation of the laser beam can be interrupted by placing an obstacle along its path. The information that the signal has been interrupted will arrive at the screen with a delay determined by the speed of light, in perfect agreement with the principles of relativity (see Fig. 4.2c). Instead, there is no way to stop the propagation of the bright spot by acting on the screen. If the obstacle is placed along the path of the bright spot, the image will go past the obstacle undisturbed (see Fig. 4.2d). Therefore, even if the spot moves faster than light, there is no superluminal communication and Einstein's theory of relativity stands firm.

In the same way, galaxies can move away at speeds greater than the speed of light without this implying superluminal communication. Galaxies move

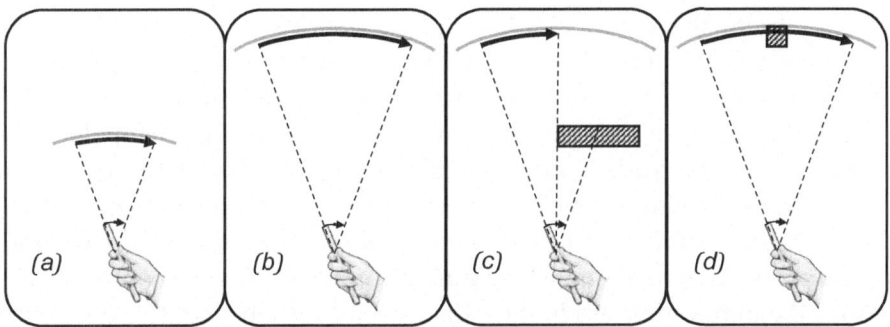

Fig. 4.2   a The beam of a laser pointer (dashed line) is directed toward a screen (gray line). By rotating the direction of the laser at a constant rate, the bright spot will cross the screen at a constant speed (black line). **b** Moving the screen further away, the speed of the bright spot will increase in direct proportion to the distance. **c** If an obstacle interrupts the propagation of the laser beam, the information will reach the screen with a delay determined by the speed of light. **d** An obstacle along the path of the bright spot on the screen will not interrupt the propagation of the signal, which only occurs along the dashed lines and not along the black line. The motion of the bright spot on the screen does not carry information and can therefore be superluminal without violating any physical law

in space at speeds lower than the speed of light, even though their distances can increase at higher speeds.

Welcome to the wonderful world of relativity.

Einstein's Conversion

Hubble's discovery of the recession of the galaxies showed physicists that the expansion of the universe was a real possibility. From the 1930s, Eddington and de Sitter took this new conception of the universe increasingly seriously and began to publicize Lemaître's results, which had remained largely unknown.

Indeed, Lemaître's 1927 article had had little influence at the time, partly because it was published in French in a Belgian academic journal with rather limited circulation. In 1931, after the discovery of the expansion of the universe, Eddington came to the rescue of his former student and friend by publishing the English translation of the article in the Monthly Notices of the Royal Astronomical Society, accompanied by a commentary written by himself. This certainly did much to bring Lemaître's scientific merits into the limelight.

This translation gave rise to a slight controversy in the history of science. The original French article contains the equation that describes Hubble's law, but it is absent from the English translation, and some modern historians have suggested that Lemaître might have been the victim of a conspiracy. Their suspicion was that an anonymous translator might have deliberately omitted the equation so as not to take anything away from Hubble's fame and glory. But the truth is more prosaic. A recent study of the correspondence between Lemaître and the editor has revealed that it was Lemaître who carried out the translation. In fact, he himself decided to leave the equation out simply because he thought everyone must already know about it.

After the publication of Hubble's data, even Einstein abandoned his skepticism. In 1932, he had a discussion with de Sitter in front of a blackboard at Caltech in California. The outcome was an article only two pages long in which they proposed an expanding universe without a cosmological constant and with just the right amount of matter to make the space flat. Einstein thus finally managed to rid himself of the creation he now most hated: the cosmological constant.

In reality, the article only describes one particular case of the solutions already set out previously by Friedmann and Lemaître. If it had not been written by Einstein and de Sitter, it would have passed completely unnoticed. But with two giants of relativity as authors, the article had resounding impact and soon converted other physicists to the idea of a dynamic universe.

A curious anecdote illustrates how the authors themselves were somehow aware of the limitations of their article. Shortly after its publication, Einstein visited Eddington in Cambridge for a few days and the two began to talk about the universe. In answer to his colleague's questions, Einstein replied: "I did not think the paper very important myself, but de Sitter was keen on it." Shortly after Einstein's departure, by sheer coincidence, de Sitter also went to visit Eddington and the conversation turned to the same topic. "You have seen the paper by Einstein and myself," interjected de Sitter. "I do not myself consider the result of much importance, but Einstein seemed to think that it was."

The expansion of the universe soon filled the pages of the newspapers and became a fashionable piece of scientific news. Hubble arranged for Einstein to visit Mount Wilson along with a swarm of journalists. Always hungry for prestige and celebrity, Hubble even invited the young film director Frank Capra (who made the unforgettable *It Happened One Night* and *It's a Wonderful Life*). In this way, Hubble could be filmed while posing alongside Einstein, who had to pretend to look through the telescope. Hubble described the meeting as an important scientific event. Einstein's diary recalls the visit as a pleasant tourist trip.

Have We Seen the Big Bang?

Despite the astronomical evidence in favor of the expansion of the universe, there was still widespread reluctance to accept the idea of the Big Bang. There was a general ideological resistance to such a conceptual revolution. Although Hubble's discovery showed that we are currently living in an evolutionary phase of the universe, this was not considered incontrovertible proof that the universe must have had a beginning. There were also doubts about the data, because the method used to derive the distances of the galaxies was open to measurement errors and the recession speeds were contaminated by local galactic motions.

But another, more worrying objection had arisen. As explained earlier, given the law of proportionality between galactic speeds and distances, the universe can be evolved backward in time to derive the moment when all the galaxies should have been located at the same point. Taking Hubble's data as input, this calculation predicted that the age of the universe was about one billion years. Yet it was already known in the 1930s from the dating of radioactive elements that the age of the Earth had to be at least two or three billion years. So, how could the Earth be older than the universe itself? Despite the attempts of some cosmologists, led by de Sitter, to find ways around this contradiction, the enigma remained unresolved and constituted a serious obstacle to believing in the Big Bang.

The explanation for this apparent contradiction was only found toward the end of the 1950s. The problem lay simply in the inaccuracy of Hubble's astronomical measurements, which did not take into account the existence of two different populations of Cepheid variable stars. This was discovered only later by Walter Baade. Current data show that Hubble's law is correct and that the Big Bang dates back to 13.8 billion years ago. The age of the Earth, on the other hand, is 4.5 billion years, so there is no contradiction. Unfortunately, the physicists of the time could not have known all this. Something else was therefore needed to convince the scientific community to accept the Big Bang.

5

The Cosmic Forge

If one tells the truth, one is sure,
sooner or later, to be found out.
Oscar Wilde

In his workshop in the innermost recesses of Etna, the god Vulcan threw stones full of iron, silver, and gold onto the fire. Under the powerful blows of his hammer, these metals were then transformed into wonderful objects, like the thrones of the gods and Jupiter's arrows. Angered by the repeated infidelities of his beautiful wife Venus, the lame Vulcan beat the metals with such violence that his blows still resonate in Etna's eruptions.

The transmutation of the chemical elements has always exerted a mysterious fascination, inspiring alchemists to go out in search of the so-called philosopher's stone so that they could transform base metals into gold. Despite its aura of esoteric mysticism, alchemy nevertheless attracted philosophers, churchmen, nobles, and, last but not least, scientists. The most surprising example must surely be Isaac Newton, one of the founding fathers of modern physics. An examination of his writings shows that he devoted a good deal of time to alchemy, even though he maintained a certain secrecy regarding these studies.

In theory, there is nothing wrong with the alchemists' dream of transforming lead into gold. But in practice, they never stood a chance. The various chemical elements differ in the number of protons contained in their corresponding atomic nuclei. To transform one element into another, protons must be added or removed from the relevant atoms, and this is no easy matter because protons and neutrons are very tightly bound in atomic nuclei. There

G. F. Giudice, *Before the Big Bang*, Copernicus Books,
https://doi.org/10.1007/978-3-031-69933-7_5

are two ways to transform elements. The first is to break the nucleus by bombarding it with projectiles made up of high-energy particles. This process, called fission, is used in the reactors of nuclear power plants. Fission can also occur spontaneously, but only in certain isotopes or radioactive elements that are rather rare in nature. The second way is to somehow force two nuclei together and thereby fuse them into a heavier one. This process is called nuclear fusion. Today, many hope it will solve humanity's energy problems. Neither approach was available to medieval alchemists, because both require a considerable amount of energy to trigger the given process, far more than they had access to. Nature, however, can do better than alchemists.

In nature, processes capable of transmuting elements can occur in high-density systems with temperatures of at least ten million degrees. Such extreme conditions are reached in the centers of stars like the Sun. In the 1930s, the pioneering studies of Hans Bethe showed that nuclear fusion is indeed the source of energy that powers stars, thus solving the long-standing enigma of how the Sun keeps us warm.

The Big Bang provides us with another physical situation in which nuclear reactions could have transformed elements: the primordial universe. Indeed, retracing the cosmic evolution backward in time, space is compressed. As anyone who has used a bicycle pump well knows, compressing a gas increases its temperature. Therefore, the universe must have been hotter in the past. Going far enough back in time, one should eventually encounter the extreme temperatures need to trigger nuclear transformation processes.

In this way, the Big Bang hypothesis provides suitable conditions for a cosmic forge in which to shape at least some of the chemical elements of the universe. The physicist who did most to promote this astonishing idea was Gamow.

Alpha, Beta, and, Above All, Gamma

Georgiy Antonovich Gamov, later known in the West as George Gamow, was born in 1904 in Odessa, a city which was in those days prosperous, cosmopolitan, and rich in intellectual life. As he recounts in his amusing autobiography, Gamow understood from childhood that his destiny was to become a scientist. It all started with an experiment carried out with the microscope his father had just given him. During Communion, instead of swallowing the piece of bread dipped in wine, he took it home to study it under the microscope. He wanted to compare it with another piece of bread, prepared earlier, in order to detect traces of transubstantiation, and thereby provide experimental verification of a dogma put out by the Russian Orthodox Church. The experiment was a failure because the young Gamow could see no difference between the two pieces of bread. However, it served to consecrate him to science.

The world war and the revolution radically changed life in Odessa. Food and water were scarce, but Gamow describes those difficult years with the sense of humor that characterized him throughout his life. Just one anecdote will make the point. As a student at the University of Odessa, he attended the multidimensional geometry lectures given in the late afternoon by Professor Kagan. Due to the shortage of fuel, the electricity supply was often interrupted, whereupon the classroom would be plunged into sudden darkness. Quite unperturbed, the professor would continue his lecture in the dark, pointing out that the various shapes of multidimensional geometry could not be seen on a two-dimensional blackboard anyway. At the end of the year, when the students passed the exam brilliantly, the professor commented jokingly: "This proves that imagination is vastly superior to illumination."

In 1923, his father sold what remained of the household silver to send young George to study in Petrograd, under the supervision of the very same Alexander Friedmann who had anticipated the Big Bang. Unfortunately, Friedmann died shortly after Gamow's arrival, and the young student turned his interests towards the emerging field of quantum mechanics.

Gamow had the opportunity to travel, spending periods in Gottingen, Copenhagen, and Cambridge, which were at the time the main centers of activity for the development of quantum mechanics. He had established a name for himself in the world of nuclear physics for having brilliantly solved an enigma concerning radioactive decay. He returned to the Soviet Union with great honors, but the country was undergoing a rapid transformation under the rise of Stalinism.

His frank and irreverent spirit was ill-suited to a scientific environment where party ideologues could decide which concepts of physics were acceptable to dialectical materialism and which should instead be forbidden. One of his public lectures was unexpectedly interrupted when he mentioned Heisenberg's uncertainty principle. He was told in no uncertain terms that such a principle was contrary to Marxist philosophy. With his usual irony, Gamow commented: "With very few exceptions, philosophers do not know much science and do not understand it. [...] But while in the free countries philosophers are quite harmless, in dictatorial countries they constitute a great danger for the development of science."

Gamow realized that this implied a considerable threat to his freedom of thought and that it might be difficult for him to obtain a passport to travel abroad. He thus decided to defect, even though he was deeply attached to his homeland.

The escape plan was a perfect mixture of ingenuity and naivety, characteristics typical of Gamow and common among theoretical physicists. After carefully studying the borders of the Soviet Union in the atlas, he decided upon what he called the Crimean campaign. He began to buy eggs on the black market, which he hard-boiled and put away as a reserve. He added a few bars of chocolate and two bottles of liquor. With the help of some friends at the sports center, he managed to buy a collapsible canoe and found a place in a holiday home for government employees that was conveniently located in Crimea. His young wife, whose nickname was Rho, was not admitted in the holiday home, but managed to rent a room in a Tatar peasant cottage not far away. The plan was to cross the Black Sea by canoe and reach Turkey. Once there, in order to avoid being identified as Russian fugitives and handed a guaranteed one-way ticket to Siberia, Gamow planned to show the authorities an old Danish driving license that he had left over from his brief stay in Copenhagen. The idea was that he would seek refuge at the Danish embassy. Then, a short phone call to his friend Niels Bohr, the famous physicist from Copenhagen, would get him out of trouble. It was a perfect plan, at least in Gamow's mind.

Having arrived in Crimea and made some trial runs in the canoe, Gamow and Rho set off in secret, rowing south with their scant provisions aboard the canoe. At the last moment, they bought a few strawberries at the local market and, after some debate, Gamow gave in to his wife, who also wanted to take along a toothbrush. The first day everything went smoothly and they rowed at a good pace. But on the second day a storm broke out and, while Gamow tried to keep control of the canoe among the violent waves, Rho was busily emptying the water from the boat with her hands. On the third day, they began to have hallucinations and at that point Gamow decided to head back. They reached the shores of Crimea, not far from where they had started, and there they collapsed exhausted on the beach. The next morning, Tatar fishermen found them semi-conscious on the sand and took them to hospital. Gamow told the authorities that they had got lost during an outing. The police believed the lie. Anyway, even the most suspicious Soviet agent would have found the truth too improbable to be believed.

Gamow finally managed to leave the Soviet Union in a much less adventurous way. He was invited to the Solvay Congress by Bohr. The physicist Paul Langevin, who was a member of the French Communist Party, acted as an intermediary with the Soviet authorities to get temporary passports for the couple. Neither Bohr nor Langevin had realized that Gamow intended to leave the Soviet Union and they were rather embarrassed when they found out, since they were the ones who had dealt directly with the authorities in Moscow. But with the mediation of Marie Curie, who was a close friend of Langevin's, the physicists finally agreed to support Gamow's decision. Rutherford, the discoverer of the atomic nucleus, and Bohr lent the couple of expatriates the money to buy tickets on a ship bound for New York.

And so it was that, in the summer of 1934, a brilliant thirty-year-old arrived in America, jovial and with a keen sense of humor, six foot three, with blonde hair and shrewd blue eyes behind glasses with lenses as thick as the bottom of a bottle. Gamow's fame had gone before him and he was immediately able to settle at George Washington University. It was here, a few years later, that he began to study the production of chemical elements in a hot universe, an area in which he became one of the pioneers.

In 1948, he wrote up the studies he had carried out in collaboration with his student Ralph Alpher. Now, Gamow was someone who could never resist an opportunity for a joke. On this occasion, he signed the paper with the names Alpher, Bethe, and Gamow, including among the authors a dear friend of his, that same Hans Bethe who had carried out the first calculations on nuclear fusion in stars, but without even warning him. He had done it purely because he liked its resonance with the first three letters of the Greek alphabet. The article was sent to the appropriate journal and, as chance would have it, it ended up on Bethe's very own desk, given his role as editor. Bethe thought "it was a rather nice joke and the paper had a chance to be correct," so he left his name on the list of authors. It was published on April Fool's Day, and subsequently became widely known under the name of $\alpha\beta\gamma$.

That same year, the young physicist Robert Herman joined Gamow's research group and Gamow insisted that he change his name to Delter. Although Herman never agreed to this, Gamow referred to a paper written with his two students as Alpher, Bethe, Gamow, and Delter. His wife Rho later recounted that Gamow "had never been happier than when perpetuating a practical joke."

The $\alpha\beta\gamma$ paper paved the way for what is today considered a cornerstone of the Big Bang theory, namely, the formation of chemical elements as a result of thermonuclear reactions in the primordial universe, a process known today as *nucleosynthesis*.

To understand the idea of nucleosynthesis, imagine being a spectator of the cosmic awakening, as the universe just begins to stir. Suppose you could take a photograph of the universe just a tenth of a second after the Big Bang. The snapshot would show a uniform gas of particles at a temperature of thirty

billion degrees. This almost unthinkable temperature would shake the particles into such a frenzy that no stable atom could possibly exist. Even atomic
nuclei would immediately be shattered by the violent collisions with other
particles.

But let's take a closer look at the photograph and make a more careful
analysis. You will see protons and neutrons (the particles that make up atomic
nuclei), with a slight excess of one over the other: for every neutron, there
will be 1.6 protons. However, protons and neutrons are drastically in the
minority compared with other particles. For every neutron, there will be 1.3
billion electrons and almost as many positrons (the antimatter counterpart
of the electron), 3.8 billion neutrinos and antineutrinos (extremely elusive
particles that are almost invisible experimentally), and 1.7 billion photons
(the particles that make up electromagnetic radiation).

Wait a moment though: how could we possibly know exactly how many of
these particles there were if no one ever actually photographed the universe a
tenth of a second after the Big Bang? This question leads us to an even more
general question: how can science even begin to investigate the Big Bang?
This question is so central to our story that a digression will be necessary.

The Laws of Physics

The scientific method is based on logical deduction from the observation of
natural phenomena or from experiments that reproduce those phenomena
under controlled conditions. In what sense then can we use the scientific
method to investigate the origin of the universe? The Big Bang is not an
observable phenomenon in nature, and neither can it be reproduced in the
laboratory. Does this mean that our understanding of the origin of the
universe is just some kind of fairy tale, devoid of scientific foundations?

Fortunately, we have a secret weapon to hand that allows us to tackle such
questions: the existence in nature of universal physical laws. This is what
allows us to embark on this incredible journey through time, and to get right
back to the beginnings of cosmic history.

Physical laws are expressed by mathematical equations that capture the
regularities observed in natural phenomena and the inexorable relations of
cause and effect that determine how a system changes. They constitute the
organizing principle that governs the dynamics of the universe and everything contained within it. They reflect a rational order inherent in nature,
which does not operate in a capricious or random way, but according to
a strict logical scheme. Physical laws are not a social construct. They hold

sway everywhere in the universe. They held sway before dinosaurs roamed the Earth, and they will still do so when no trace of humanity remains.

The deep reason for the existence of an ordered system of laws underlying natural phenomena remains a mystery, perhaps one that sits beyond the reach of scientific inquiry. However, the existence of these laws is not a philosophical hypothesis or a religious dogma, but a well-established empirical result. And in fact, the better we understand nature, the clearer the unified image of physical laws that emerge, involving increasingly complex mathematical structures and ever simpler basic principles. Physical laws provide the key to understanding the universe and embody an inexplicable and undeserved gift that nature has given to humanity. If there were no physical laws in nature, the scientific method itself would serve no purpose.

Experience teaches us that there is a world of objective reality, independent of our perception. Physics teaches us that there is also a world of abstract forms, populated by absolute truths that can be demonstrated using mathematical logic. This world is reminiscent of the *eidos*, the Platonic idea of perfect forms. The extraordinary achievement of science has been to discover a relationship between two aspects of nature that at first glance seem completely independent. Physical laws provide the link between the world of experience and the world of rationality, materializing the direct correspondence between objective reality and mathematical construction.

One key feature of physical laws is their universality, that is, the fact that the property holds, not just for the particular phenomena from which it was deduced, but in a broader setting. Universality allows us to extrapolate knowledge acquired in a certain context and apply it to new situations, even to those that cannot be reproduced experimentally. For example, using the same equations that describe the fall of an apple, Newton deduced the motion of celestial bodies, without embarking on any interplanetary journeys. Likewise, from laboratory measurements of spectral lines in the light emitted or absorbed by various chemical elements, we can deduce the composition of stars without sending any spacecraft to collect samples of stellar material. From the study of nuclear reactions, the processes that occur inside the Sun have been identified without the need to build an artificial star in the laboratory.

The universality of physical laws is what allows scientists to overcome the limitations imposed by the requirements of direct observation or experimental reproducibility. In this process of extrapolation beyond the limits of perceptibility, experimental exploration does not consist in recreating the phenomenon in the laboratory or observing it directly, but in deducing its consequences and subjecting them to empirical tests.

Universality means that physical laws are the same everywhere in the cosmos, in every place and at every time. Astronauts walking on the Moon take big steps as though bouncing across the surface, not because the law of gravity is different on the Moon, but only because the Moon is less massive than the Earth. Circumstances may change, but not physical laws.

The universality of physical laws is not absolute, but limited to a certain domain of application, and their validity must be continually called into question. For example, Newton's law of gravity works very well here on Earth or on the Moon, but it fails in the vicinity of a black hole. In situations where gravity is very strong, Newtonian theory must be replaced by Einstein's general theory of relativity. Cases like this, where a physical law ceases to be valid, do not signal a failure of the logical scheme of nature, but only the discovery of a deeper level of understanding.

The universality of physical laws provides the wings for the human mind to take flight, allowing it to travel back in time and far into space, beyond the limits of anything we can witness directly, all the way to the discovery of the Big Bang.

Cosmic Alchemy

Let us return to our imaginary snapshot of the universe a fraction of a second after the Big Bang. It is thanks to our knowledge of physical laws that we can reliably describe the cosmos, even in such extreme conditions. In particular, the relative amounts of the various kinds of particles present in the primordial gas can be deduced by assuming that thermal equilibrium is established by the incessant collisions between the particles. Once this equilibrium has been reached, the evolution of the primordial gas can be calculated quite straight-forwardly, without having to travel back to the beginning of the universe and take photographs.

As the universe expands, the temperature of the cosmic gas decreases, and neutrons and protons combine via nuclear chain reactions like those shown in Fig. 5.1, forming hydrogen, helium, and their isotopes (i.e., nuclei with the same number of protons, but different numbers of neutrons). In just over a quarter of an hour after the Big Bang, the definitive values for the mass fractions of the chemical elements would have been reached: 75.5% hydrogen, 24.5% helium, 0.003% deuterium, 0.002% helium-3, and traces of lithium.

At first glance, this prediction may not look very accurate. On Earth, hydrogen is combined with other chemical elements, but the atmosphere

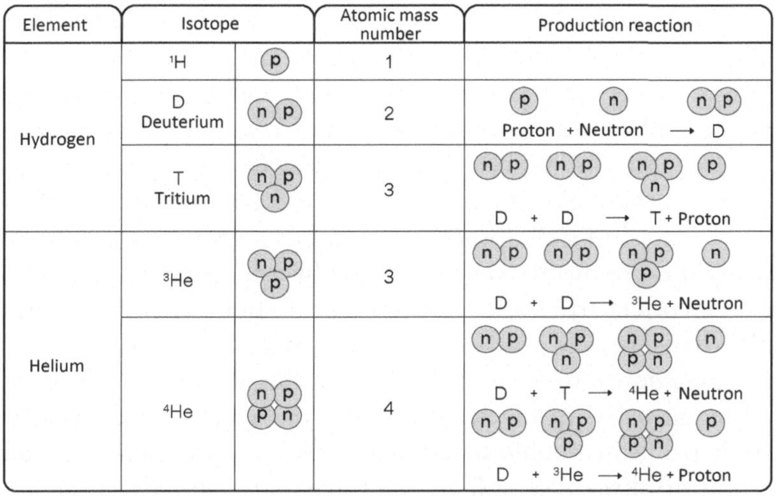

Fig. 5.1 The main reactions contributing to the cosmic nucleosynthesis of light chemical elements. Isotopes are elements whose atomic nuclei contain the same number of protons, but a different number of neutrons. The atomic mass number is the total number of protons and neutrons in the nucleus

contains only 0.5 millionths of pure hydrogen and 5 millionths of helium (by mass). However, we should not let this lead us astray. The chemical composition of the Earth and its atmosphere are not representative examples of the general characteristics of the cosmos. In particular, hydrogen and helium are very light gases and they easily escape the Earth's gravitational attraction, so they are very scarce on our planet. Astronomical measurements of the chemical composition of stars and interstellar material, on the other hand, clearly demonstrate an excellent agreement with the amounts derived from the most modern calculations of nucleosynthesis.

Gamow laid the conceptual foundations of primordial nucleosynthesis, but the details of his analysis were not correct. The authors of the $\alpha\beta\gamma$ paper made the hypothesis that the universe had started out as a gas of neutrons, which they called *ylem*. This is a word from Middle English which Alpher had stumbled across in the dictionary. It was used in medieval philosophical writings to speak of a primordial substance. Gamow liked the word so much that he later convinced himself that he had actually invented it. According to $\alpha\beta\gamma$, the ylem produced all the chemical elements through decays of neutrons into protons and subsequent capture of free neutrons. In reality, the chain of nuclear reactions involved in nucleosynthesis is much more complex, bringing together a great many other processes.

Gamow knew his analysis was incomplete, but he did not have all the tools required to carry out a more systematic calculation. Not all the reaction rates of nuclear processes were known at the time, and many that were known could not be exploited because they concerned research on the atomic bomb and were thus treated as military secrets. Despite Gamow's study being largely wrong, it was nevertheless of great importance because it showed that the Big Bang was not just a philosophical idea, but a genuine scientific hypothesis that could be verified or indeed refuted by astronomical observations. In physics, an incorrect article that contains a profound idea may be much more influential than a correct article that contains only bland ideas.

In 1956, Gamow moved to the University of Colorado, where he divorced Rho and remarried with his publisher. He was in fact an original popular science writer. The incredible adventures of the shy bank clerk Mr. Tompkins in the world of physics are still an enjoyable and instructive read. Unfortunately, his abuse of alcohol took its toll on his health, and it also marred his reputation when he appeared at conferences in a state of inebriation, interrupting the speaker with incoherent questions or falling asleep noisily. He died at sixty-four of cirrhosis of the liver. The physicist Edward Teller remembered him thus: "Gamow was fantastic in his ideas. He was right, he was wrong. More often wrong than right. Always interesting [...] and when his idea was not wrong, it was not just right, it was new."

The study of nucleosynthesis has built up a firm connection between the microworld and the cosmos. The physical laws that govern the reactions between some of the smallest components of matter—namely atomic nuclei—also govern the global structure of the universe. This connection between microworld and cosmos has become ever more intricate as research progresses, and now underpins our study of the universe. In this way, elementary particle physics and cosmology have become inextricably interrelated areas of research, each drawing motivation from the other.

It is fascinating to know that the light chemical elements in the universe are the product of nuclear reactions that took place in the first few minutes after the Big Bang. This is an overwhelming piece of evidence in favor of the Big Bang theory, because cosmic nucleosynthesis could only have been triggered if, at some point in its past history, the universe had reached temperatures of at least a few tens of billions of degrees, about a thousand times hotter than the center of the Sun.

When it was proposed, however, the idea of primordial nucleosynthesis was still incomplete and insufficient to convince the scientific community. The existence of the Big Bang remained an open question.

6

The Eternal Universe Strikes Back

*There is no mistake so great
as that of being always right.*
Samuel Butler

It may seem strange to us today, but many physicists found the Big Bang hard to swallow. Perhaps there were still echoes of Einstein's blunt appraisal: "The idea makes no sense." Eddington in an influential paper of 1931, hence after Hubble's discovery, wrote: "Philosophically, the notion of a beginning of the present order of Nature is repugnant to me."

For many, it was hard to abandon the solemn eternity of the universe in favor of something that sounded too much like a biblical tale. After the Second World War, it had been accepted that the universe was in an evolutionary phase with an expanding space, as suggested by astronomical observations. However, the hypothesis of an initial Big Bang was still not taken seriously. Gamow and Lemaître were exceptions. So, it should be no surprise that the physics community was ready to welcome a new proposal for an eternal universe with open arms.

In 1948, three Cambridge physicists, Fred Hoyle, Hermann Bondi, and Thomas Gold, put forward the idea of a stationary, or steady-state universe. Rather than 'static,' which basically means that nothing is happening, the term 'steady state' is used in physics to speak of a system with its own internal motion, but whose overall properties do not change as time goes by, like a steadily flowing river. For example, a gas in a container kept at constant temperature and pressure is in a steady state. The molecules of the gas are

© The Author(s), under exclusive license to Springer Nature
Switzerland AG 2024
G. F. Giudice, *Before the Big Bang*, Copernicus Books,
https://doi.org/10.1007/978-3-031-69933-7_6

moving all the time, but viewed as a whole, the system does not appear to change.

According to Hoyle, the idea of the steady-state universe was born when the three friends went together to see the film *Dead of Night*. This was one of the first horror films produced in Britain after a ban, during the Second World War, on any kind of show that might lower people's morale.

The film begins with the main character, the architect Walter Craig, arriving at a cottage in the English countryside, where he is to oversee a renovation project. Upon his arrival, Craig confesses to his hosts that he has already met them in a series of recurring nightmares, despite the fact that he has never actually seen them before. Moreover, it seems that he is able to foresee events in the cottage before they happen. His hosts decide to put him to the test and each tells him a story of supernatural events. Shocked by the frightening stories, Craig attacks one of the hosts and tries to strangle him, but then he wakes up suddenly, realizing that it was all a dream. In the morning, he receives a phone call for a renovation job and off he goes to the very same cottage where the film started.

As they left the cinema, Gold asked his two companions: "How if the universe is constructed like that?" The universe might unfold like the plot of the film, with lots of things happening, but nothing actually changing.

The idea was to build a universe that was stationary rather than static. A universe which is always the same, without beginning and without end, which is uniform not only in space, but also in time. But how could they reconcile this idea with astronomical observations of the expansion of space? According to Hoyle, Bondi, and Gold, this was possible thanks to a perpetual creation of matter from nothing, which could fill the new space generated by the expansion and maintain the matter density constant everywhere. Clearly, Hoyle, Bondi, and Gold's universe was not static, because space was expanding in accordance with astronomical measurements, but it was in a steady state, because the accompanying creation of matter meant that the global characteristics of the universe were always the same as time went by.

How is it possible to reconcile the creation of matter from nothing with the familiar notion of energy conservation? Hoyle and his colleagues calculated that, to keep the universe in its steady state, it would be enough to create just a few atoms per cubic kilometer of space and per century. The injection of new matter would thus be so tiny that it could easily escape experimental detection. While in the Big Bang, all matter is produced in a single instant, in the steady-state universe matter is produced eternally, but only in very small amounts.

Viewed through modern eyes, the theory looks contrived. Yet, the philosophical attraction of a theory that did not require a cosmic beginning was so great that Hoyle, Bondi, and Gold's steady-state universe was received with great enthusiasm. And so, a scientific controversy was born between these two schools of thought: the Big Bang versus the steady-state universe.

It was indeed a colorful controversy due to the personalities involved. There was an interesting parallel between the contenders. Gamow's histrionics were counterbalanced by Hoyle's equally pronounced exuberance. Ambitious, brilliant, and eclectic, Hoyle had excellent communication skills. His science fiction stories had considerable success with the general public. I still remember how impressed I was as a boy watching the television drama *A for Andromeda*, which was based on one of his stories. It tells of a group of scientists who use information picked up from Andromeda to build a supercomputer capable of creating an alien life form. There are also secret agents and espionage ... more than enough to send a teenager's imagination into orbit.

Notwithstanding the similarities between the two contenders, there were differences in their personalities. Gamow's humor came in the form of off-the-cuff remarks and witticisms, the kind that could trigger laughter all round. Hoyle's humor, on the other hand, was sharper and more cynical. Indeed, he was prone to sarcasm and sometimes came across as fiercely provocative.

Gamow's two young collaborators, Alpher and Herman, were both the sons of Russian Jews who had emigrated to the United States. While the former had experience in applied physics, the latter was a relativist. Similarly, Hoyle's collaborators, Bondi and Gold, were both Viennese Jews who had met in an internment camp in Quebec in 1940. While Bondi was inclined toward mathematics and relativity, Gold was interested in applied problems and had worked on the physical mechanism underlying the functioning of the inner ear.

The exuberant Gamow and Hoyle loved to pursue the contest even outside the academic environment and duly extended it to the general public. It was during a BBC radio program that Hoyle invented the term 'Big Bang.' His intention was to poke fun at the theory and make it seem absurd. Instead, he inadvertently guaranteed its media success. The expression 'Big Bang' was soon common parlance and an established part of scientific jargon.

When the debate first began, the wind seemed to be blowing in favor of the steady-state universe. Hoyle's reputation among astronomers carried considerable weight, but the main issue was the age of the universe. As explained earlier, the astronomical data available at the time suggested that the universe born from the Big Bang had to be younger than the Earth, a paradoxical result to say the least. Clearly, there was no such contradiction with a steady-state universe, which was, by definition, eternal.

The confrontation between the Big Bang theory and the steady-state universe may seem like a rather abstract matter. However, there was a perfectly good way to find out which was right: to identify the origins of what we ourselves are made from.

Are We Children of the Stars or Relics of the Big Bang?

Gamow had a brilliant intuition: the chemical elements we observe today in the universe might have been synthesized in thermonuclear reactions that occurred during a very hot primordial phase. This is precisely the process of nucleosynthesis, which explains the cosmic abundances of hydrogen, helium, and their isotopes. Despite this first success, Gamow had encountered a hitch.

In nature, there are no stable chemical elements with atomic mass number 5, that is, with a total of 5 protons and neutrons in the atomic nucleus. This purely chance fact blocks the chain of nuclear reactions that can occur in the primordial universe, preventing the efficient production of chemical elements heavier than helium (with the exception of a very small trace of lithium). In the cosmic forge, the chemical elements are produced in sequence. The fact that there are no stable nuclei of mass 5 blocks this line of production at mass 4, and so is unable to produce anything of mass 6 or beyond.

Gamow was convinced that this was a temporary problem, and that it was just a matter of identifying the right nuclear reaction. Enrico Fermi also became interested in the problem, but neither he nor others were able to overcome the obstacle at mass number 5. So, for the moment, the cosmic forge was able to produce the correct quantities of hydrogen and helium, but could not go beyond. However, our bodies are also made of oxygen, nitrogen, carbon, calcium, phosphorus, and so on. Where do these elements come from?

For Hoyle, this difficulty spelt failure for the Big Bang hypothesis. Naturally, throughout its unchanging eternity, the steady-state universe could never have been any hotter than today, so there was no way it could ever have accommodated cosmic nucleosynthesis. Hoyle thus put forward the idea that all the chemical elements we find in the universe today were produced from free protons and neutrons in the nuclear reactors that naturally exist within stars. The confrontation between the Big Bang theory and the steady-state theory could therefore be resolved by addressing the question of the origin of the chemical elements: were they produced in a remote hot phase of the universe or in the centers of stars?

A crucial turning point came with the discovery that the simultaneous fusion of three helium nuclei can efficiently generate carbon (with mass number 12) in a hot environment, provided that the matter is dense enough. These conditions are realized in the center of stars, but not in the cosmic forge, because matter would have become too rarefied by the time the universe reached temperatures suitable for these nuclear reactions. Put another way, the blockage at atomic mass number 5 is easily overcome in stars, where heavy chemical elements can be produced, while it remains insurmountable for cosmic nucleosynthesis.

By now, the wind was blowing even more strongly in favor of the steady-state universe. However, it was soon found that the fraction of hydrogen that converts into helium in stars was too small to explain the great abundance of helium found in the universe as a whole by astronomical observation. So, who was right? Are the chemical elements remnants of the Big Bang or are they produced by stellar processes?

In his autobiography, Hoyle explains with his usual incisiveness and irony how scientific disagreements are resolved: "Of course, the prisoner in the dock knows already whether he is guilty or not. In court, the prisoner hopes the jury gets it right if he knows he's innocent and he hopes the jury gets it wrong if he knows he's guilty. In physics, on the other hand, the jury of experimentalists can be taken always to be right. The problem is that you don't know whether you are innocent or guilty, which is what you stand there waiting to learn as the foreman of the jury gets up to speak."

The verdict regarding the question of the origin of the chemical elements was … that the two contenders were both right and wrong at the same time. Nature loves to conceal the simplicity of its principles and manifest itself in a complex way. The mistake that physicists had made was to believe that all the chemical elements had to have the same origin. Instead, it is known today that different mechanisms come into play, some cosmological and some astrophysical.

The lightest chemical elements, like the isotopes of hydrogen and helium, were indeed produced in the cosmic forge shortly after the Big Bang. Intermediate elements, such as beryllium or boron, can be produced by cosmic rays when they collide with and fragment heavier nuclei. On the other hand, all the chemical elements we are most familiar with, such as the oxygen in the air, the carbon in our cells, or the silicon in rocks, are produced inside stellar forges and then scattered throughout the universe by gigantic explosions called supernovas. So, we are indeed stardust. Heavy chemical elements, such as gold, platinum, and uranium, can be produced in accidental collisions between neutron stars. These are extremely dense celestial bodies, with a mass equal to that of the Sun concentrated in a radius of just a few kilometers. The existence of these processes was confirmed by the detection of gravitational waves from such a collision in 2017. Finally, all the elements heavier than plutonium are synthesized in the laboratory and do not exist naturally.

We may conclude by paraphrasing a quip by Gamow, who noted that the light chemical elements were baked by the universe in less time than it takes to cook a dish of roast duck and potatoes. But it subsequently took billions of years to recook the resulting hydrogen and helium inside the stars and produce the other chemical elements necessary for our existence. In short, the recipe for life requires a very long preparation time.

In the end, there is no single explanation for the origin of all the chemical elements. As it turns out, the decisive evidence for the Big Bang did not come from the origin of the elements, but from another sensational discovery: observation of light emitted by the Big Bang.

In Praise of Oblivion

Before discussing how the light left over from the Big Bang was able to shed light on the question of the Big Bang, it will be useful to make a small digression and ask what we can learn from the steady-state theory. Despite its popularity between the late 1940s and early 1960s, there is no trace of this theory in modern cosmology textbooks. It survives only in books about the history of physics. This is as it should be because, in science, it is the experimental data that determine the fate of theories and the data have condemned the steady-state universe to oblivion.

The story of the steady-state universe is a lesson on the way ideological considerations can lead scientific research astray. Hoyle was ideologically opposed to the Big Bang because of its possible religious interpretations. He thus did everything he could to devise an alternative theory. The result looks contrived today, and much less convincing from a conceptual point of view than the Big Bang theory.

The steady-state universe shows that scientific progress is not a straight path, but a twisted tangle of brilliant ideas and wrong turns, of fundamental discoveries and blunders. Mistakes are not necessarily failings. They are often necessary in order to arrive at the truth. As the philosopher of science Karl

Popper insisted: "To avoid error is a poor ideal: if we do not dare to tackle problems which are so difficult that error is almost unavoidable, there will be no growth in knowledge. In fact, it is from our boldest theories, including those which are erroneous, that we learn most."

Research is all about having ambitious goals and getting carried away by our intuitions and enthusiasm, even at the risk of running at full speed into a brick wall. Only when pursuing mediocre research can one know in advance what results will be obtained. Great scientific ideas cannot be planned, and rarely arise out of the blue. They are often the fruit of stubborn failed attempts and great mistakes. Every physicist knows only too well that the wastepaper basket is an indispensable tool for their work. It is where the majority of ideas will eventually end up. Nonetheless, these discarded ideas will serve to fertilize the ground, giving rise perhaps to that rare and isolated intuition that drives scientific progress.

This uncompromising but necessary process of selection is common to all creative activities, not just scientific ones. Artists know this only too well, and many of them remain perpetually dissatisfied with their own works, repeating the same subject again and again in their obsessive search for perfection. Michelangelo, filled with frustration, pounded his *Pietà Bandini* (the Deposition) with a hammer, disfiguring the body of Christ. Leonardo da Vinci left many paintings unfinished, although they are today celebrated as masterpieces. There is a crucial difference, though. In art, the value of a work is decided by the subjective taste of the author or some grumpy critic. In science, the final verdict on the validity of a theory is always left to experiment. We leave it entirely up to nature to decide whether a scientific idea is right or wrong.

Another consequence of the steady-state universe—and herein lies its most important contribution—was to attract the attention of the scientific community. Up until the late 1940s, cosmology had been a peripheral scientific activity, practiced by few and even then only sporadically. Most physicists looked on cosmology with a certain distrust, treating it more like a form of divination than a branch of science. Although it was generally accepted that the universe was in an evolutionary phase of expansion, the idea that it might have had a beginning seemed to belong to the field of metaphysics.

By offering a concrete alternative to the Big Bang, the steady-state universe forced physicists and astronomers to consider the question in scientific terms. The controversy between the Big Bang theory and the steady-state universe—at least within the academic environment—revolved around the calculation of nuclear reactions and comparisons with astronomical data, not on ideological positions. In the end, it was the experiments and not the philosophical assumptions that determined the winner. But the experiment that conclusively consecrated the Big Bang was very different from what any of the contenders had imagined.

7

Light from the Big Bang

Research is what I'm doing when I don't know what I'm doing.
Wernher von Braun

Science has a rich tradition of great discoveries made purely by chance. In 1895, Wilhelm Röntgen, while working in the darkness of his room on experiments with a cathode ray tube covered by a thick piece of black cardboard, was surprised to see a green light appear on a distant fluorescent screen. Using his apparatus, he took a photograph of his wife's hand. Horrified at seeing an image of her bones, she exclaimed: "I have seen my death!" And so it was that X-rays were discovered.

In 1896, Henri Becquerel, annoyed by the bad weather that prevented him from continuing his experiments on the phosphorescence induced by sunlight, stored away in a dark cupboard his apparatus based on uranium salts. Despite the darkness, an image remained imprinted on the photographic plate. And so it was that radioactivity was discovered.

In 1928, Alexander Fleming interrupted his studies on staphylococci to go on holiday, forgetting an open container. Upon his return, he found it covered with a greenish mold that had killed these pathogenic bacteria. And so it was that penicillin was discovered.

The list of such accidental discoveries is long. Horace Walpole, 4th Earl of Orford, invented a word to describe a discovery made with a mixture of luck and unconscious intuition: serendipity. The name was inspired by the book *The Three Princes of Serendip*, written by Cristoforo Armeno in 1557, which recounts the imagined adventures of the three princes of the kingdom of Serendip (present-day Sri Lanka). One of the most sensational cases of

G. F. Giudice, *Before the Big Bang*, Copernicus Books, https://doi.org/10.1007/978-3-031-69933-7_7

serendipity in the history of science was the experiment that finally ensured the triumph of the Big Bang theory.

How the Princes of Serendip Discovered the Light from the Big Bang

Cyanogen is a compound of carbon and nitrogen, normally occurring in the form of a highly toxic transparent gas with an unpleasant smell. Astronomers first detected it in the tails of comets, a discovery that caused panic in 1910, on the eve of the passage of Halley's Comet near the Earth. Even without the internet to fuel fake news, the population was gripped by the fear that the comet's tail would release a lethal cloud of cyanogen into the Earth's atmosphere. In New York, gas masks, anti-comet pills, and even protective umbrellas were sold, and they clearly worked because no victims were reported.

Cyanogen is also found in interstellar space. The chemical substances present in the cosmos can be identified by analyzing the light that passes through them. In fact, each substance absorbs the particular frequencies of light that stimulate the quantum transitions between different energy levels of the electrons in the atoms. Spectrographic analysis of the light that passes through a gas shows dark lines at these frequencies, thus revealing a kind of fingerprint that can be used to identify the different chemical substances in the gas.

The surprise was that interstellar cyanogen contained not only the spectral line corresponding to the transition from the lowest energy level, but also a second line corresponding to an excited state of the gas (hence at a higher energy level). So, why is there so much excited cyanogen in interstellar space?

In 1940, the Canadian astronomer Andrew McKellar ventured a guess. If the cosmos were filled with thermal radiation, interstellar cyanogen would be in a state of equilibrium between two energy levels, constantly absorbing and re-emitting radiation at a well-defined frequency. From astronomical data, McKellar calculated that the temperature of this hypothetical cosmic radiation would be about 2.3° above absolute zero. The result was curious enough, but not many paid heed to it. McKellar's premature death in 1960 after a long illness certainly did nothing to make his prophetic discovery better known in the academic world.

During their pioneering studies on nucleosynthesis, Gamow, Alpher, and Herman understood that a very hot initial universe must necessarily contain intense electromagnetic radiation that would have survived until today, albeit greatly attenuated by the expansion of space. In 1948, Alpher and Herman, unaware of McKellar's result, calculated that the residual radiation must today have a temperature of 5° absolute. Given the general skepticism regarding the Big Bang, the result did not attract much attention and even the authors themselves never believed that such weak radiation was observable. The moral is that scientists either take themselves too seriously or not seriously enough, but rarely find the right middle ground.

Hoyle—the champion of the steady-state universe—knew of McKellar's result and, in 1956, while Gamow was taking him for a ride in his white Cadillac, they discussed it together. Gamow shook his head and replied that the cosmic radiation from the Big Bang should have a much higher temperature than had been deduced by McKellar. So, the conversation on the subject came to a halt, just like the Cadillac when it came to a red light. Later, Hoyle would even cite the measurements of interstellar cyanogen as evidence against the Big Bang.

By this time, Alpher and Herman had already left the academic world to work in industry, while the eclectic Gamow had turned his interests to the structure of DNA. But our own story now comes to the decisive turning point.

In 1963, Arno Penzias and Robert Wilson, two radio astronomers employed by Bell Telephone Laboratories in the United States, began work on converting a telecommunications antenna into a radio telescope. Unfortunately, the antenna seemed to have a defect. No matter how much they tried, they were unable to eliminate an annoying background hum in the microwave frequency spectrum. They tried everything, even using a shotgun to get rid of some pigeons that were leaving droppings on the antenna. But the mysterious hum would not go away.

Meanwhile, less than fifty kilometers away from the Bell Laboratories, another key actor entered the scene: the physicist Robert Dicke, then at Princeton University. Using the experience on radar he had acquired during the Second World War, he had the idea of scanning the sky at low frequencies, in search of a signal coming from the cosmos. In 1964, he asked two of his students, Peter Roll and David Wilkinson, to help him build the apparatus, and enlisted a young theoretical physicist, Jim Peebles, to calculate the spectrum of electromagnetic radiation that should reach us from the depths of cosmic history.

The two radio astronomers at the Bell Laboratories were completely unaware of all this activity in nearby Princeton. Quite fortuitously, during a conversation with an old friend, it was suggested that Penzias get in touch with Professor Dicke to see if he might shed light on the origin of the problem with the antenna. So, Penzias picked up the phone and called Princeton.

At that exact moment, Roll, Wilkinson, and Peebles were in Dicke's office for a lunch break centering on a few sandwiches and, as a side dish, a relaxed physics discussion. The phone rang and Dicke answered. Even as Penzias was finishing his story, Dicke already understood that someone had inadvertently pipped them to the post. He lowered the handset for a moment and whispered to his young collaborators: "Well, boys, I think we've been scooped!".

And so it was that, in 1965, the Princeton group went to visit Penzias and Wilson's antenna, to discuss their results with them. A fair compromise was reached and two papers were published, one immediately after the other. In the first, the duo at Bell Laboratories described how they had measured an unknown microwave radiation, distributed evenly across the sky. In the second, the Princeton group interpreted the measurement as a signal from the earliest days of the universe. The new discovery was called the *cosmic microwave background radiation*, recording the fact that it had initially manifested itself as an unwanted background noise.

The cosmic background is a spectacular example of an accidental discovery. However, as almost always happens, behind the serendipity, there was serious scientific work. There was the foresight of the Bell Laboratories in offering the possibility of independent research to its employees, in the hope of deriving benefits for applications in radio communications. There was the stubborn determination of the researchers to fully understand their tool and perfect it to the point of discovering even something they were not looking for. And there was the theorists' intuition that allowed them to understand the meaning of the data and see the dawn of the universe behind what appeared to be a mere background noise.

This cosmic radiation thus became the irrefutable proof that the universe is not in a stationary state but was once very different from today. Not only is space expanding, but in the early days the universe must have been much hotter, and it must have reached very high temperatures. If, when we go back in time, the space of the universe contracts and its temperature increases, at some point in cosmic history, we must come to a Big Bang. With the discovery of the background radiation, we stumbled across the light left by the aftermath of the Big Bang. Penzias and Wilson were awarded the Nobel Prize in 1978, and Peebles in 2019.

What is the Cosmic Background Radiation?

Light (or, to be more precise, electromagnetic radiation of any frequency, in the visible spectrum, in radio waves, or in gamma rays) travels at around a billion kilometers per hour. It's a huge speed, but not infinite. For this reason, we see astronomical bodies in the sky as they were in the past.

We see the Sun as it was eight minutes ago. We see Antares, the reddish star in the center of Scorpio, as it was at the time when Leonardo da Vinci was learning his trade. We see Andromeda, the galaxy closest to us, as it was when our *Australopithecus* ancestors were wandering the plains of Africa. Astronomers observe the center of the Markarian 231 galaxy, the nearest quasar to us, as it was when the first complex multicellular organisms were evolving on Earth. The further away the object we observe, the further back into the past its image takes us. Telescopes are time machines.

When we observe the night sky, we see the universe projected onto the surface of the celestial vault, like a fresco painted on the ceiling of a medieval cathedral, lacking any perspective. The depths in the image, or the distances of the celestial bodies, are just not represented. The celestial vault completely flattens all depth onto a single surface, thereby superimposing many images of the universe that correspond to enormously different times. It's as if all the pages of a history book were printed on a single sheet.

All we can observe of the four-dimensional cosmic space–time is its two-dimensional projection cast upon the sky. The signals that reach us from the depths of the universe must first be decoded to reconstruct the distances to the stellar bodies that sent them, and hence untangle the web of overlapping images on the celestial vault. Only then can we read the history of the universe directly from what we see in the sky (see Fig. 7.1).

In our vicinity, there are planets and stars. Beyond, there are galaxies similar to our own Milky Way. Pushing on further, there are younger galaxies, some still forming. And then there is darkness, because the universe was filled with a rather uniform gas of hydrogen and helium and, being transparent, this is invisible to telescopes. But the story doesn't end there.

Moving toward greater distances, and therefore more remote times on the journey back to the Big Bang, the temperature of the transparent gas increases due to the contraction of the universe. We thus reach a point where the temperature of the hydrogen and helium gas is so high that collisions due to thermal motion are capable of tearing electrons from the atoms, thereby ionizing the gas. At this distance, there is a luminous surface, called the last scattering surface. It is a glowing plasma that emits intense electromagnetic radiation at a temperature of 3000°, about half the temperature at the surface

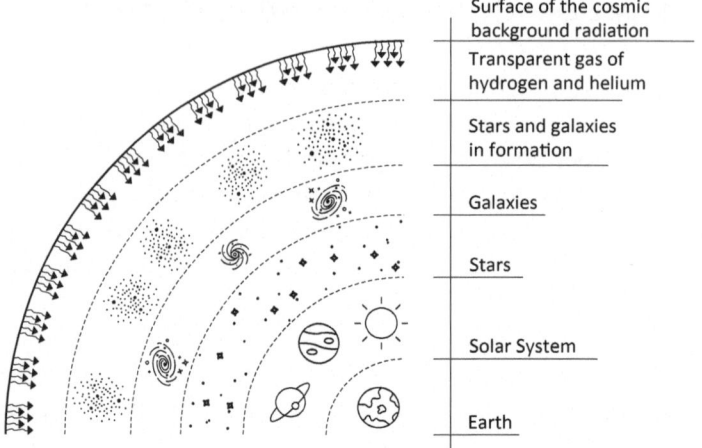

Fig. 7.1 Schematic view of the celestial vault decomposed into successive planes, corresponding to ever greater distance scales, or ever more remote times

of the Sun. Although very hot, the plasma is rather rarefied, with a density comparable to the atmosphere at the surface of the Moon. This plasma forms a gigantic spherical luminous screen that surrounds the universe and effectively releases a snapshot of the cosmos taken 380,000 years after the Big Bang. It is this plasma screen that emits the cosmic background radiation.

Here on Earth, we observe the image of this screen after its light has traveled through the cosmos for 13.8 billion years. The expansion of space distorts the original image, stretching the wavelengths of the radiation, so that it arrives here at a temperature about a thousand times lower than when it was emitted. That's why on Earth we observe only weak radiation at a temperature of 2.725° above absolute zero (about 270° below zero Celsius) and with frequencies mainly in the microwave spectrum, just like those produced in kitchen appliances.

The fact that we are exactly at the center of the sphere described by the cosmic radiation should not lead us to believe that we live at the center of the universe. For any alien anywhere in the universe would see exactly the same thing. It's like being in the middle of the ocean and noticing that the horizon describes a circle centered on our own location. If we were anywhere else on the ocean, we would have the same impression.

The plasma behind the last scattering surface teems with free electric charges to the point of being impenetrable to any kind of light. This surface marks the boundary between a transparent gas of atoms and a completely opaque plasma made up of atomic nuclei and electrons moving too fast to bind together. This surface forms a luminous background wall to the universe, beyond which optical observations are no longer possible. We cannot see what lies behind.

Astronomy thus finds itself confined to a prison with insurmountable walls, corresponding to the epoch 380,000 years after the Big Bang. Nothing that happened before can be observed with measurements of electromagnetic radiation (radio waves, microwaves, infrared, visible light, ultraviolet, X-rays, or gamma rays). In particular, the event corresponding to the Big Bang itself is confined to darkness behind the wall of cosmic radiation, and we will never be able to see it directly.

Even so, cosmological exploration pushes well beyond that wall today. For example, the nuclear fusion processes that gave rise to the light chemical elements occur behind the plasma screen of the background radiation. In this case, we found a way past the screen using a wise blend of theoretical calculations, indirect observation, and logical deduction. Nuclear physics was the key tool for understanding nucleosynthesis, but if we want to get even closer to the Big Bang, we need to deepen our knowledge of particle physics.

Although the wall emitting the cosmic background radiation blocks the passage of any electromagnetic radiation, there is something else that can get through it: gravitational waves. In 2016, a major announcement was made by the LIGO project based in the United States, in collaboration with the Virgo project located in Italy: the first direct observation of gravitational waves, produced by a merger of two black holes. The result was rewarded with the 2017 Nobel Prize. Experimental programs for observing gravitational waves are undergoing enormous development at the moment, and in the near future, it may be possible to study gravitational signals originating behind the wall. To put this poetically, it is sometimes said that astronomers have until now only observed the light of the universe (i.e., electromagnetic waves), while tomorrow we will be able to listen to the sound of the universe (i.e., gravitational waves).

Nonetheless, there is no reason to torment ourselves over the difficulties of seeing through the last scattering surface, because the image emitted from it is already an inexhaustible mine of information about the primordial universe. Soon after the discovery of the cosmic background, several projects were devised to obtain ever more precise measurements. The first goal was to measure the radiation at different wavelengths to reconstruct its entire thermal spectrum. To begin with, this was done with experiments aboard atmospheric balloons and sounding rockets, but it soon became necessary to place a detector in Earth orbit in order to completely eliminate the absorption of radiation by the upper atmosphere.

In 1990, NASA's COBE satellite made measurements across the entire celestial sphere and over a wide range of wavelengths, from microns to centimeters, providing an extraordinary confirmation of the thermal nature of the cosmic background radiation. The data demonstrated beyond any reasonable doubt that the temperature of this radiation is nearly uniform in every direction of the sky, thereby proving that the universe emerged from the Big Bang in a state of general uniformity throughout space. The two principal investigators of the COBE mission, John Mather and George Smoot, were awarded the Nobel Prize in 2006.

It is curious to note that the temperature of the cosmic background deduced by McKellar from the spectral lines of cyanogen in 1940 was, within the measurement errors of the time, in substantial agreement with the value of 2.725° above absolute zero, established today with great accuracy. As further proof, in 1993, new measurements were made of excited cosmic cyanogen, and these confirm the perfect agreement. So, McKellar was right. It was only due to a piece of bad luck that no one listened to him at the time, and it took another twenty-five years before the scientific community became aware of the existence of cosmic radiation.

Cosmic Time

The extraordinary uniformity of the temperature of the cosmic background shows that the pioneers of cosmology made the right hypothesis: to a first approximation, the universe is indeed uniform in space. At night, we see only a dark sky dotted with stars because of the evolutionary accident that made our eyes sensitive to the frequencies of light in which the Sun is brightest. An alien whose eyes were sensitive only to frequencies around a hundred gigahertz would see a diffuse and uniform light across the whole sky at any time of day or night. This is the cosmic background radiation.

The uniformity of cosmic space is not just a useful tool that allows us to simplify Einstein's equation. It is also the key to addressing an essential issue if we want to talk about the history of the universe: the problem of measuring cosmic time.

Einstein taught us that time is not absolute, but depends on the position and motion of whoever measures it. An alien living near a neutron star observes time to pass much more slowly than an earthling. Two astronauts who meet in cosmic space, speeding past each other in opposite directions, will not even agree on which of two flashes of light arrived first. In short, Einstein tells us that time is a subjective matter.

What sense is there then to talk about a flow of time valid throughout the whole universe? What does it mean to say that the age of the universe from the Big Bang to today is 13.8 billion years? Relative to which alien, astronaut, or earthling have 13.8 billion years gone by?

Time is measured by change: the sand flowing in an hourglass, the hand that turns on a clock, or the Earth that rotates on its polar axis and orbits around the Sun. Cosmic time is measured by the evolution of the universe: the drop in temperature of the primordial gas, nucleosynthesis, the formation of atoms, and so on. Cosmic uniformity allows us to define a universal clock that is valid everywhere in the universe, since the entire cosmos follows the same evolutionary history, provided that observations are made at large enough distance scales, ignoring local inhomogeneities. This clock is determined by the general structure of the universe and not by the particular position or speed of some alien.

Before the discovery of cosmic radiation, the uniformity of the universe was only a theoretical hypothesis. Now, it has become an incontrovertible fact. If the universe were not globally uniform, cosmology would be a chaotic discipline. We would not even be able to give an absolute meaning to the temporal sequence of cosmic events. Cosmology would become the study, not of the history, but of the sociology of the universe. It is extremely lucky for professional cosmologists that this is not the case, and that it is possible to tell the story of the universe in an orderly manner and give a precise meaning to a statement like this: the Big Bang happened 13.8 billion years ago.

The Sound of Light

After the success of COBE, the exploration of the cosmic background radiation continued with other experiments, including NASA's WMAP satellite and the European Space Agency's Planck mission, which operated between 2009 and 2013. The measurements have allowed us to reconstruct a very accurate map of the cosmic background, and new missions are being planned. But what is the point of such an intense observational program?

Measurements of the cosmic background radiation have taught us that the universe was exceptionally uniform in its infancy. Yet, it could not have been perfectly uniform. This is because the planets, stars, galaxies, and all the structures that are observed today in the universe attest to the existence of inhomogeneities in the distribution of primordial matter, inhomogeneities that were then amplified by the attractive effect of gravity. If at the beginning the universe had been perfectly uniform, gravity would never have been able to amplify anything, and we would just be atoms of a homogeneous gas today.

This means that, just after the Big Bang, there must have been variations in the density of matter, and these variations would have left their mark on the cosmic background radiation. The purpose of the WMAP and Planck space missions was to study the structure of these characteristic marks.

The primordial radiation behaves like a gas of photons (the particles that make up electromagnetic radiation), just as the air behaves like a gas of molecules. The cosmic gas of photons is subjected to two opposing forces. On the one hand, gravity attracts the photons toward regions where matter is denser. On the other hand, there is a pressure that opposes condensation of the photons and tends to repel them, making their distribution more uniform.

The two opposing tendencies thus create oscillations in the gas, with the photons trying to fall toward the points of gravitational attraction, then bouncing back, driven by the pressure. These compression and rarefaction oscillations in the photon gas are exactly analogous to sound waves propagating in the air. In other words, the primordial radiation is expected to be full of acoustic waves.

The surface emitting the background radiation carries an image of the acoustic waves that were propagating in the universe 380,000 years after the Big Bang. This image of the sound of cosmic light was first observed by the COBE mission in 1992 and is shown in Fig. 7.2 in the more accurate version obtained by the Planck space telescope. The figure is effectively a map of the whole sky, where the shades of gray represent variations around the average temperature.

A weather map, even of a relatively small country like Italy, shows significant temperature variations between different cities. It would be no surprise to find that, on the same day of the year, the temperature was 5 °C in Bolzano and 20 °C in Palermo. Now, the map in Fig. 7.2 covers a region much larger than Italy. In fact, it covers the entire observable universe, spanning distances up to tens of billions of light-years. And on this huge map, the temperature of the cosmic radiation varies by at most one part in a hundred thousand, attesting to its extraordinary uniformity. At one point on the celestial vault, the temperature might be 2.72545° absolute and at another 2.72551, but the differences are never greater. So, on the face of things, the cosmic weather report looks terribly boring. However, for a cosmologist, these tiny temperature variations are absolutely thrilling because they contain truly precious information about the primordial universe. To give just one example: on this map, we can read the geometry of cosmic space. Let's see how.

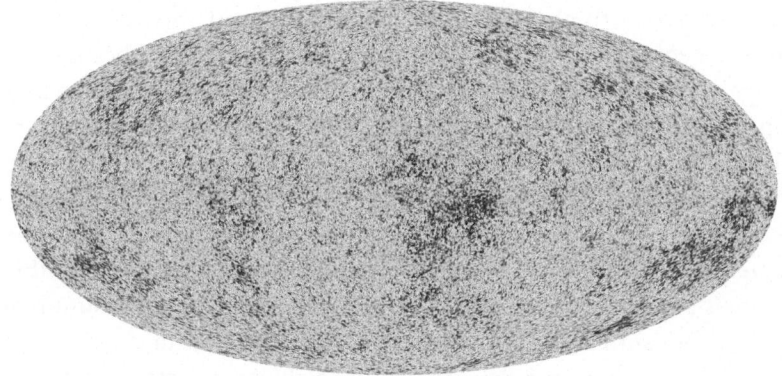

Fig. 7.2 Celestial map of the temperature variations in the cosmic background radiation as measured by the Planck mission of the European Space Agency. The average temperature of the radiation is 2.725° above absolute zero, while the different shades of gray indicate temperature variations, which are at most a few tens of millionths of a degree

Imagine a paleontologist who stumbles upon an exceptional discovery. During an excavation in a remote village, she finds the bones of a new species of dinosaur: the strangeosaurus. The villagers are very excited by the discovery and eager to communicate the information to the rest of humanity. In their world, however, there is no radio, telephone, or internet, so they can only communicate directly from person to person. They set off in different directions to tell anyone they meet the amazing story of the strangeosaurus. Some of them walk quickly, others run, but no one can move faster than ten kilometers an hour. It is clear that, after an hour, the news of the strangeosaurus can have traveled at most ten kilometers, but no further. In other words, no person living beyond ten kilometers from the village can have received the information about the strangeosaurus, an hour after its discovery.

Something similar happens for the spotty pattern on the map in Fig. 7.2. Different points of any given spot must have had time to exchange information to reach the same temperature. Since Einstein's relativity teaches us that no physical information can be transmitted faster than light, the spots on the celestial map cannot be larger than the distance traveled by light in the 380,000 years that have elapsed since the Big Bang. In other words, by a simple calculation, we can deduce the maximum extent of the spots, just as we deduce the maximum extent of the region within which the existence of the strangeosaurus could be known. What is interesting now is to compare this theoretical calculation with astronomical observations.

Our spaceborne instruments do not, however, produce a true image of these spots in the cosmic radiation. Instead, they measure the apparent image created by the deformation that the light rays undergo on their journey through the geometry of space. The reason why there is such a deformation is once again the unusual property of non-Euclidean geometries encountered in Chap. 2, according to which the sum of the angles of a triangle can differ from 180°. As explained in Fig. 7.3, in a closed universe, the spots that describe the temperature variations appear to the Planck instruments larger than their actual size. In an open universe, the spots appear smaller than their actual size. In a flat universe, there is no deformation. This effect is illustrated in the panels at the bottom of Fig. 7.3, which show numerical simulations of the deformation suffered by the image of cosmic radiation due to the geometry of space. The three panels correspond to the same actual spotty pattern, but are seen through spaces with different curvatures.

The deformation of images in a curved space is analogous to viewing objects through optical lenses. A convex lens makes light rays converge, enlarging the true image, while a concave lens has the opposite effect (see Fig. 7.4). Living in spaces with non-Euclidean geometry is like seeing the world through distorting lenses. However, there is an important difference. The distortion of the path of light by lenses only occurs at their surface, while in a curved universe, the distortion occurs at every point in space.

By comparing the theoretical calculation of the true image with the experimental measurement of the apparent image, we can determine the curvature of space. The data from the Planck mission indicate that the universe has the flat Euclidean geometry on these very large scales, at least to within an experimental uncertainty of less than one percent. Naturally, this does not mean that the geometry of space is flat everywhere. In the vicinity of a black hole or a neutron star, space–time is highly curved. What these measurements of the cosmic background radiation show is the extraordinary flatness of the overall structure of the geometry of the universe.

It is truly astonishing how humanity has managed to determine the shape of the cosmos up to the most remote distances, combining deductive reasoning with extraordinarily precise measurements made by space missions. The result is a remarkable triumph of the scientific method.

At the same time, it is a triumph that comes with a big surprise. Among all the possible geometries that the overall structure of cosmic space could have had, nature has chosen a particular form: Euclidean geometry. It may seem reassuring that the universe as a whole follows the familiar rules of geometry discovered 2300 years ago by Euclid, and not the weird contortions of non-Euclidean geometry. But, as we will see in Chap. 8, this result presents us with an enigma.

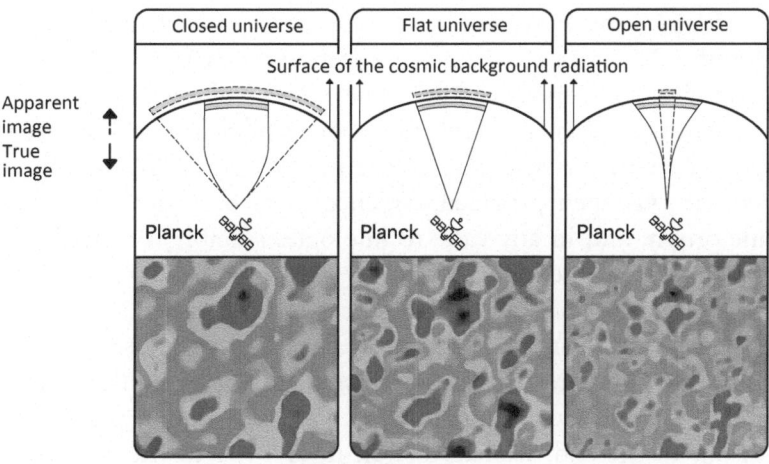

Fig. 7.3 A curved universe deforms the image of the spots that characterize the temperature variations of the cosmic background radiation. The true image of each of these spots corresponds to the shorter side of the triangle constructed with two rays of light that reach the Planck space observatory. In a curved space, the sum of the angles of a triangle is not 180°. For this reason, Planck sees an apparent image of the spots larger than the true image in a closed universe, and smaller in an open universe. Only in a flat universe do the apparent and true images coincide. The panels at the bottom show numerical simulations of the map of temperature variations in the cosmic radiation for the three different types of space geometry. From the comparison between these simulations and the data in Fig. 7.2, a measure of the geometry of the universe can be extracted

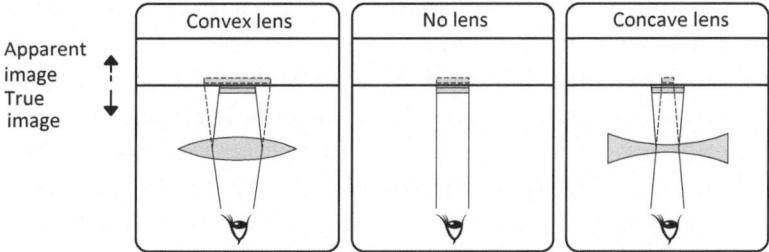

Fig. 7.4 Optical lenses shrink or enlarge images in an analogous way to the curvature of space

Why Did It Take so Long to Make This Discovery?

The story of the discovery of the Big Bang is so full of missteps that one might wonder why the journey was so tortuous and why it took so long to reach a conclusion. After all, the requisite theoretical knowledge and astronomical data were available well before the scientific community finally accepted the idea of a progenitor event. The general theory of relativity, essential for describing the shape of space–time, was completed in 1915. The recession of the galaxies was confirmed in 1929. The nuclear physics needed to understand the synthesis of chemical elements was sufficiently well understood as early as 1932, with the discovery of the neutron. In 1940, the observation of interstellar cyanogen provided evidence in favor of a thermal radiation of cosmic origin, and in any case, technologies suitable for the detection of microwave electromagnetic radiation were accessible shortly after the Second World War. However, no one put the pieces together.

Progress was slow, on both the theoretical and the observational front, especially if we compare it to the explosive pace of new ideas and discoveries in the fields of quantum mechanics and subatomic physics over the same period. In the end, we had to wait until 1965 to witness the ultimate success of the Big Bang theory, with a discovery made by chance: the cosmic background radiation.

There can be no doubt that one cause of this apathy was a deep philosophical resistance to the idea of a cosmic beginning, mixed with a general skepticism about whether cosmology could ever make significant scientific progress. This is also the reason why it is impossible to identify a single author of the Big Bang theory. Several personalities contributed significantly to the development of ideas before the discovery of the cosmic background radiation, but none of them had the vision, the know-how, or the perseverance to put together all the elements and build a complete theory.

Sometimes making a scientific discovery does not involve uncovering something radically new, but simply seeing with different eyes what is under everyone's nose. Scientific ingenuity is not only expressed through pioneering discoveries, but also through the ability to sense the value of information that is already available. Sometimes, it's just about observing the pieces of the puzzle that are already on the table, recognizing their significance, and finding the right way to put them together. This too is the result of genius.

Science moves through a tangled web of ideas, and scientific progress never follows a straight path. Scientists are rarely completely free from their ideological prejudices, which sometimes help them to have the right intuition, and sometimes prevent them from recognizing what the data are telling them. However, the scientific method is such that, in the end, the truth will always emerge. As Victor Hugo wrote, "No army can resist an idea whose time has come." When the time is ripe, ideas prevail over prejudices. So it was for the Big Bang.

Despite the familiarity of the term 'Big Bang' in everyday language today, we should not forget just how revolutionary it is. The preconception that the universe must be static and eternal, without beginning or end, was so deeply rooted in physics that the very idea of cosmic evolution seemed alien to science. The initial distrust of the Big Bang and the subsequent stubborn resistance by scientists in the first half of the twentieth century testify to its enormous intellectual import. This discovery has forced us to say goodbye to cosmic immutability, depriving us of a reassuring antidote against the precariousness of human existence. Here, we are faced with an immense conceptual revolution.

The Birth of a New Science

The monumental discovery of the cosmic background radiation cut the Gordian knot on the origin of the universe, pronouncing in favor of the Big Bang. But there's more. The discovery also marks a turning point for cosmology, which has been transformed from a predominantly speculative and descriptive activity to a quantitative science, where calculations based on established theories are compared with precise astronomical observations. In short, with this discovery, cosmology shed its carefree youth and adopted a conscious maturity.

Before the discovery of the cosmic background, cosmology was an activity confined to the imagination of a few physicists. The best minds were engaged elsewhere. Quantum mechanics had undermined the principles of classical physics, opening the way to mind-blowing discoveries regarding the intimate structure of matter and new technologies that have since revolutionized our everyday lives. These were the scientific problems that clamored for attention. The Big Bang was just a vague concept beyond the boundaries of knowledge, or worse, a temptation to be resisted.

The discovery of the cosmic background changed the course of science. The Big Bang theory has become the paradigm for addressing specific questions about the origin of the universe, which until then had remained under the sole dominion of religion or philosophy. Cosmology thus set off along a path of increasing rewards, gaining its well-deserved recognition in the scientific community.

8

The Enigma of the Big Bang

The universe is like a safe to which there is a combination.
But the combination is locked up in the safe.
Peter de Vries

In a survey conducted in 2014, 51% of Americans stated that they did not believe in the Big Bang, or at least considered the theory to have little credibility. After giving a lecture in an American high school, a friend told me that, faced with the question: "What is the Big Bang theory?" the majority of students answered: "A TV series."

What is the Big Bang Theory?

The Big Bang theory is a rigorous scientific theory based on the hypothesis that, at a certain moment in the past, the universe was made up of a rapidly expanding space and a hot, dense, and almost perfectly uniform mixture of particles, although with a few small variations in its density. Given this hypothesis about the initial conditions of the universe, it is then entirely up to the laws of physics to determine the evolution of the system and the destiny of the cosmos.

The logical structure of the Big Bang theory rests on three extremely well-established pillars. The first is general relativity, which describes the geometry of space and its expansion. The second is particle physics, which describes the quantum properties of matter and radiation. The third is statistical mechanics, which describes the behavior of high-temperature physical

G. F. Giudice, *Before the Big Bang*, Copernicus Books, https://doi.org/10.1007/978-3-031-69933-7_8

systems. Starting from these three pillars of knowledge, now confirmed by countless laboratory experiments, the theory is able to make very precise predictions which can then be compared with astronomical observations.

The credibility of the theory is based on three empirical observations. The first is the recession of the galaxies, which reveals the expansion of space and therefore, according to Einstein's equation, the existence of an initial event hidden in the past. The second is the great abundance of helium in the cosmos and the excellent agreement between astronomical observations and predictions of nucleosynthesis. This result shows that the temperature of the universe was extremely high in the past. The third is the observation of the microwave background radiation, evidence that the universe was not only hot in the past, but also almost perfectly uniform. In the face of these experimental data, not believing in the Big Bang is like claiming that the Earth is flat.

The Big Bang theory is one of the most extraordinary achievements of human understanding, as it provides us with a quantitative description of the history of the universe. At the same time, it gives us a splendid synthesis of the physical world, bringing together the laws that govern the microscopic world and the global properties of the cosmos in the most surprising way.

Clearly, we do not yet know everything about the history of the universe after the Big Bang. There are still many unanswered questions, especially regarding the origin and nature of certain components of the universe, viz., dark energy and dark matter, but also regarding the prevalence of matter over antimatter. These are certainly fascinating questions for current research in cosmology, but they fall outside the scope of this book. In these pages, we shall instead attempt to understand what caused those special initial conditions of the universe that we call the Big Bang.

The Big Bang and the Beginning of Everything

At this point, we should clarify what exactly we mean by the Big Bang. When the equations of relativity are used to rewind the evolution of the universe back in time, we come to what physicists call a *singularity*. By this, we mean a point where space–time is infinitely curved and the laws of general relativity are no longer applicable. According to Einstein's equation, 13.8 billion years ago, our universe would have encountered a singularity, where all the distances between points of space were zero and matter was infinitely dense and infinitely hot.

No one should believe that the universe was ever really in this paradoxical situation. The result only indicates that some new physical phenomenon must necessarily intervene before space collapses into the singularity. Indeed, we know that Einstein's general theory of relativity fails when we approach such a singularity because it cannot describe the extreme physical conditions that are reached in those situations. Before encountering the singularity, the universe enters a regime where the curvature of space–time is so high that it will no longer be governed by general relativity. In this regime, we do not yet have the physical understanding needed to give an adequate description of cosmic evolution.

Astronomical data does not tell us that the universe began with a singularity, but only that there must have been a moment when the cosmic matter was in an extremely dense and hot state. There is no reason to believe that the temperature or the density of matter was infinite at that moment. We only know that, in order to ignite the process of nucleosynthesis, the temperature of the primordial universe had to be at least one hundred billion degrees, that is, thousands of times higher than in the center of the Sun. It is quite possible that it was enormously higher, perhaps even hundreds of billions of billions of billions of degrees, but not infinite. As we will see later, we may one day be able to determine the initial temperature of the universe, because its measurement may be within reach of new experiments to study the cosmic background radiation.

In light of the astronomical data, I will define the Big Bang as *the event that gave rise to a universe made up of a nearly uniform mixture of particles at high temperature and density, in an expanding space.* It is the initial moment of a hot primordial soup that contains all the ingredients of today's universe.

My definition does not refer to any singularity of space–time. The term 'Big Bang' is sometimes used to refer to the event that gave rise to space–time. At that instant, all reality takes shape and nothing could have existed before. To avoid confusing the two different definitions, I will call the hypothetical moment of cosmic creation the *Beginning of Everything*.

It is worth bearing in mind that the Big Bang and the Beginning of Everything are in principle two distinct events, and there is as yet no experimental data to suggest that they must coincide. The first corresponds to a well-determined physical phenomenon and marks the beginning of a phase in which the universe is describable by currently known physical laws. Its existence is attested by many astronomical observations that determine its characteristics. The second is only the result of a conjecture, and there is no way of checking it at the present time since it refers to extreme conditions that cannot be described by the theories we have at our disposal today.

Moreover, we have no empirical evidence of the existence of a Beginning of Everything, nor reliable clues to help us. So, we cannot say for certain whether the universe ever experienced a Beginning of Everything.

While we can say with great accuracy when the Big Bang occurred, there is no reason to believe that the Beginning of Everything must have occurred 13.8 billion years ago, even if it really existed. For the moment, the Beginning of Everything is just a vague abstract idea. Many of the theoretical proposals regarding the origin of the universe do not refer in any way to an initial moment when space–time suddenly materialized and became a physical reality.

For the time being, we shall therefore concentrate on understanding the Big Bang. Later, in Chap. 15, we shall make a quick foray into even more remote times, in search of a hypothetical Beginning of Everything.

The Mysteries of the Big Bang

The Big Bang theory does not attempt to say what exactly the Big Bang is. It only describes what happened after that fateful event, and it is completely silent about the mechanisms that may have produced it or indeed anything that happened before. For the theory, that event is only a hypothesis about the initial state of the universe, and not a specified physical phenomenon. The Big Bang theory is not a theory *of* the Big Bang.

The hypothesis on which the theory is based—that the universe began in a nearly uniform, hot, dense, and expanding state—might seem a reasonable and innocent assertion. However, it hides deep mysteries. There is nothing generic about it at all, and some of its features are so remarkable as to make it almost inconceivable. These are the mysteries that will guide us in the search for the origin of the Big Bang.

This situation is reminiscent of *The Murders in the Rue Morgue* by Edgar Allan Poe, where the eccentric Auguste Dupin has found an opportunity to exercise his exceptional deductive abilities by investigating a double murder committed in a Paris apartment. The police are at a loss. The clues are so implausible that it seems the crimes could not possibly have been committed by a human being. But starting from these apparently incomprehensible clues and using only pure reasoning, Dupin does of course discover the truth. We are in a similar situation when it comes to the origin of the Big Bang: the clues are mysterious and the solution will surpass all imagination.

Mystery Number 1: The Expansion

The Big Bang theory hypothesizes that the universe began in a phase of spatial expansion. But what physical phenomenon could have given the initial push that got that expansion going? What was the engine that triggered the Big Bang?

The Big Bang was not an explosion that occurred at a point in space. If it were, we would be able to locate the center of the explosion in the cosmic background radiation, or some trace of the shock wave that would have followed. Instead, the cosmic radiation is almost perfectly uniform and does not manifest any preferential direction. This indicates that the Big Bang occurred almost simultaneously throughout space. Therefore, it is an event localized in time, not in space.

It is no easy matter to identify a mechanism that could have started the expansion because, among all the fundamental forces we know of, only the gravitational force is capable of giving rise to effects that extend to great cosmic distances. This suggests that the explanation must somehow lie in gravity. Yet Newton taught us that gravity is a purely attractive force. So, it can cause a contraction, or at best slow down a pre-existing expansion, but it is not capable of triggering an expansion.

Mystery Number 1: How could the Big Bang act simultaneously across the whole of space, initiating its expansion?

Mystery Number 2: The Uniformity

The Big Bang theory hypothesizes that the universe began in a nearly uniform state, as corroborated by observation of the cosmic background radiation. Yet, this uniformity is itself a puzzle.

The temperature of the cosmic background radiation has the same value at every point in the sky, with deviations only in the fifth decimal place. At first glance, there is nothing particularly strange about this. If you leave a cup of hot coffee in the kitchen and come back much later, the coffee will have cooled, equalizing its temperature with that of the air. But the point is that it takes time to reach thermal equilibrium and the universe did not have enough time. It was too young at the moment when the image of the background radiation was formed, 380,000 years after the Big Bang.

Places in the universe that we observe today at opposite directions in the sky were, at the time when the background radiation was emitted, too far apart to have ever exchanged any physical information, given that, according to Einstein's theory of relativity, such information cannot propagate faster than light. How is it possible that places in the universe that have never been in contact and have never been able to communicate with each other share the same temperature? This is as paradoxical as finding that the hot coffee left in the kitchen mysteriously reaches the same temperature as all the cups of coffee left by unknown aliens in their kitchens on distant planets.

Mystery Number 2: How could the Big Bang apparently fail to comply with the laws of relativity and make the universe almost perfectly uniform, even at points in space that should, according to those laws, never have been able to communicate with each other?

Mystery Number 3: The Flatness

As we saw in Chap. 7, the measurement of the cosmic background radiation shows that the geometry of the universe is almost perfectly flat. At first glance, this may not seem to be a problem because one might think that any geometry is as good as any other. But, as soon as we take into account the expansion of the universe, this too presents us with a mystery.

Any tiny deviation from flatness will grow wildly during cosmic evolution. Just as celestial structures collapse under the effect of gravity, so the geometry of a curved space becomes increasingly curved as cosmic time goes by. For the universe to be as flat today as inferred from observational data, its flatness must have held to an accuracy of one part in a billion billion at the time when nucleosynthesis occurred and with a dizzyingly greater accuracy at times even closer to the Big Bang. Flatness is a singularly unstable condition under cosmic evolution.

As an analogy, imagine that the position of a marble placed on the pyramid of Cheops indicates the value of the curvature of the universe. The lower the marble is, the more the geometry is curved. When the marble is exactly on the top of the pyramid, the geometry is flat.

Now imagine that someone (the Big Bang) selects the initial position of the marble, that is, the initial condition for the geometry of the universe. You didn't get there in time to witness the initial event but, 13.8 billion years later, you go to Egypt to see where the marble is. Naturally, you would expect to find it somewhere near the bottom, or at least on its way down. Instead, you find it still very close to the top. You deduce that, at the beginning, the marble must have been positioned remarkably close to the apex of the pyramid, with absolutely extraordinary accuracy.

There is no logical inconsistency in what you saw in Egypt. It could just have been a coincidence. However, in science, we are wary of coincidences on the verge of the impossible, for they may not just be the result of pure chance. They may be telling us that some phenomenon has not been properly understood. The initial condition of flatness is so special that it is natural to wonder why the Big Bang chose this particular value with such absolutely insane accuracy.

Mystery Number 3: How could the Big Bang flatten the geometry of the universe with an accuracy that is really at the limit of what seems plausible?

Mystery Number 4: The Arrow of Time

In his *Confessions*, Saint Augustine expresses the confusion we feel when we think about time: "What then is time? If no one asks me, I know what it is. If I wish to explain it to someone who asks, I do not know anymore." There is something elusive, indeed something ambiguous, about the concept of time. We perceive it as a river that always flows at the same rate, and carries the whole world along in its flow. But this intuition seems to clash with Einstein's theory of relativity, according to which space and time are manifestations of a single entity: space–time. And yet, in space, we can move freely in all directions, while time is a one-way street where it is not allowed to turn and go back. Common experience teaches us that there is an arrow of time that always inexorably indicates the same direction from the past to the future.

When we turn to the laws of physics, Saint Augustine's question becomes even more puzzling. The laws of gravity, electromagnetism, and quantum mechanics are perfectly indifferent to the direction of time. No one would notice anything strange if we reversed a film showing the oscillations of a perfect pendulum or the motion of the molecules of a gas in thermal equilibrium. The physical laws that govern these processes do not recognize the arrow of time. But, we could only laugh at the absurdity of a film in which a thousand shards of glass scattered on the floor spontaneously jumped up onto the table and reconstituted a glass. In everyday life, we immediately recognize the direction of time. The story of a man steadily getting younger until he becomes a newborn child uttering his first cries can work for a fantasy movie like *The Curious Case of Benjamin Button*, but cannot be part of reality.

Although physical laws do not generally change under time reversal, there is one that does distinguish the direction of time. This is the second law of thermodynamics, according to which the entropy of a system always increases. Translated into common language, this law expresses the idea that the disorder of a physical system (quantified by the entropy) increases as time goes by. In reality, the second law of thermodynamics is not a fundamental law of nature, but only a probabilistic accident. To understand this last claim, it will be useful to give an example.

If you keep shuffling a deck of cards initially ordered by number and suit, you can be sure that the order will gradually disappear. Why? The reason is purely probabilistic. The number of different ways to sequence fifty-two cards is colossal—as many as there are atoms in the entire galaxy—but there is only one in which the cards are ordered by number and suit. This means that reconstructing the order of the cards by continuing to shuffle the deck is, from a practical point of view, essentially impossible. It's like stubbornly looking for a particular atom hidden somewhere in the galaxy. The number of sequences of fifty-two cards is so gigantic that, at every deal, you can be practically certain that that hand has never been played before in the whole history of humanity, and will never be played again in the future of the universe. So, enjoy every game, because you can be sure it is absolutely unique.

Now, in a typical physical system, there will not be just fifty-two molecules, like the number of cards in a pack. They will be counted rather in units of Avogadro's number, which is equal to six hundred thousand billion billion. The number of possible configurations of molecules in a complex system is so colossal that it would certainly make your head spin. In short, the probability that a physical system will evolve spontaneously from disorder to order (i.e., from a state of high entropy to one of low entropy) is, for all intents and purposes, zero.

If you really want to draw a moral from a physical law, the lesson is that you should enjoy every moment of life because, as the second law of thermodynamics teaches us, that moment is truly unrepeatable.

The second law of thermodynamics is the assertion that the total number of possible states of a complex system is huge, while the ordered states are extremely rare. Hence an ordered state will nearly always evolve toward a disordered one and the opposite is so unlikely as to be effectively impossible. Our perception of the flow of time is not the consequence of any fundamental physical law. It stems from the fact that we are not elementary particles, but part of a complex system consisting of a huge number of molecules. We perceive the world by taking averages over the enormous numbers of possible configurations of the components of such systems, rather than observing the motions of individual components. The flow of time results from this approximate perception of reality, which allows us to grasp the evolution of complex systems. We sense the passage of time because the past state of the universe was less likely than the current one.

For the universe to sustain a direction of time and host an advanced evolutionary form like life, it must start from a very highly ordered state. Otherwise, if the universe was disordered at the outset, it would always be disordered, just as a deck of cards remains disordered, no matter how much we continue to shuffle it. The universe would remain forever in a sterile state of chaos, incapable of accommodating any complex evolutionary process.

The problem of the arrow of time translates into finding a condition of extreme order (hence, low entropy) for the universe at the instant of the Big Bang. Then the second law of thermodynamics will take care of degrading that order until it extinguishes the stars, destroys all life, and annihilates everything that makes the universe interesting.

Intuitively, you might expect a universe where matter and energy are uniformly mixed together to correspond to a highly disordered state. Indeed, if you pour milk into a cup of coffee and then stir it with a spoon, you transform the system from an ordered state, in which the milk and coffee are separated, into a disordered one, in which the mixture of milk and coffee is homogeneous. But alas, intuition leads us astray, because the universe is not a latte.

When we consider the large-scale structure of the universe, the only force in play is gravity, and gravity behaves in quite the opposite way to a latte. Since gravity is an attractive force, a uniform distribution of matter corresponds to an ordered state, while the entropy increases whenever matter collapses. The state of maximum disorder (at least ignoring quantum effects) is a black hole, where all the matter scattered throughout the universe gets swallowed up in a small region of space.

To explain the arrow of time, the universe must therefore emerge from the Big Bang in a state where matter and energy are as uniformly distributed as possible in space. This is the almost perfectly ordered state which minimizes the entropy and thereby allows the existence of a cosmic flow of time, rather than producing a universe that could only ever remain in a state of stationary chaos, like the well-mixed cup of latte.

The origin of the arrow of time is still a mystery. We have certainly come a long way in understanding its meaning, but we nevertheless find ourselves in a similar position to Saint Augustine, who concluded his reflections with the words: "I confess to you, Lord, that I still do not know what time is."

Mystery Number 4: How can the Big Bang produce a state of the universe that is so unbelievably highly ordered, apparently contradicting the dictates of the second law of thermodynamics?

Mystery Number 5: Cosmic Structure

The cosmic background radiation suggests that, at the moment of the Big Bang, the universe was uniform. Yet today the cosmos, at least at distances less than about three hundred million light-years, is anything but uniform. This means that the initial condition of the universe cannot have been one of perfect uniformity, but must have incorporated the density variations which eventually led to all known cosmic structures, from superclusters of galaxies to the most insignificant dwarf planets.

Mystery Number 5: How can the Big Bang imprint on the almost perfect uniformity of the primordial universe the inhomogeneities that are the origin of all cosmic structures?

The Mystery of Mysteries

The above mysteries surrounding the Big Bang might seem rather abstract issues, suitable only for discussions among theoretical physicists in the corridors of some dusty research institute. But in reality, they reflect a single, greater mystery that can be summed up in a very simple question: *was the universe created to host life?* This is the mystery of mysteries.

If the initial conditions of the universe had differed only very slightly from those hypothesized by the Big Bang theory, cosmic history would have proceeded in a radically different way. So different, in fact, that no complex structure of any kind, let alone life, could ever have come into existence. This is basically what the five mysteries listed above are telling us.

A closed universe would last only a relatively short time before curling up on itself in a catastrophic Big Crunch. An open universe, on the other hand, would expand so quickly that all the matter in it would be rapidly diluted, leaving behind an almost empty space. In both cases, there would be no time for galaxies to form, and no hope of witnessing biological evolution. In short, the emergence of life takes time and only a nearly perfectly flat universe can guarantee enough of it.

If the primordial matter had been distributed unevenly, the universe would soon have been full of black holes, ready to devour everything around them. If it had been perfectly homogeneous, the universe would have contained only a barren, uniform gas. In either case, there would be no place in the cosmos where complex structures could come into being. To end up with life somewhere in the universe, a very precise set of conditions must be fulfilled, producing a uniform mixture of matter with a light sprinkling of inhomogeneities.

The mysteries of the Big Bang, described in this chapter, tell us that the initial conditions of the universe are so special as to seem paradoxical, if not in contradiction with the laws of physics. But there's more. The equations of general relativity allow a huge variety of possible universes, but almost none of them fulfill the necessary conditions for the evolution of any form of life. From this perspective, the universe in which we live appears highly improbable. It is almost as though it has been carefully selected by an invisible hand.

The initial conditions hypothesized by the Big Bang theory are the only ones capable of producing a universe that could host biological evolution. It's as if the cosmos knew from the very beginning that its destiny was to create life. It almost seems that the universe created by the Big Bang was coded in such a way as to generate a little girl on a train, wondering whether the description of cosmic history must necessarily take into account her existence. This is the mystery of mysteries.

The many enigmas raised by the Big Bang have got us up against the wall. If we are to claim that we understand the origin of the universe, we cannot hide from these questions by simply accepting unreasonable hypotheses about the initial conditions of the cosmos. The time has come to understand the physical mechanism behind each of these mysteries and provide the solution. If there is a scientific explanation for the mysteries of the Big Bang, there is only one way to find out: we must think about what happened before. Before the Big Bang.

9

How Does the Big Bang Work?

*The people who are crazy enough to think they
can change the world are the ones who do.*
Steve Jobs

The career of a theoretical physicist usually follows more or less the same path. After obtaining a PhD, there begins a nomadic life, in which periods of two or three years are spent with fixed-term contracts called postdocs, in academic institutes scattered all around the world. Two years in some university in Great Britain, then three years in California, and another two in a research institute in Germany. In this way, physicists wander from one side of the Atlantic to the other.

It is a stimulating life, full of opportunities, in which we confront different realities and cultures, form new collaborative relationships with researchers from other countries, and get involved in fascinating projects. They are exciting years, in which we often reach the peak of our intellectual creativity. But they are also the years in which we form bonds and families, and this roaming around the globe often requires sacrifices in our personal life.

After a handful of these postdocs, the physicist arrives at a decisive moment in their career, when they may aspire to a professorship in some university or research institute. Not everyone passes this hurdle. Those who do not find the right opening are forced to look for a job outside the academic world and say goodbye to theoretical physics.

Toward the end of the 1970s, Alan Guth found himself at precisely this critical point in his career. From a Jewish family in New Jersey, Guth had married a girl he met in high school, had a son, and had four postdocs under

G. F. Giudice, *Before the Big Bang*, Copernicus Books,
https://doi.org/10.1007/978-3-031-69933-7_9

his belt in as many institutes scattered across the United States. Despite his obvious talent, things were not going particularly well for him. He had not yet produced any research result capable of opening the doors to a professorship and he had reached an age generally considered to mark the last chance for continuing in an academic career. Then, on the night of 6 to 7 December 1979, everything changed.

Guth cuts a rather unusual figure. Or perhaps I should say, he's perfectly conventional as far as theoretical physicists are concerned, because he fits the classic stereotype to a tee: absent-minded, disorganized, and absolutely brilliant. His spontaneous courtesy manifests a natural kindness, and his clarity and depth of thought reveal the greatness of his mind.

On that night in 1979, Guth developed an idea that had been buzzing around in his head for a few months. The next morning, he got on his bike and beat his own personal record on the route from his home to his office at the SLAC laboratory in Stanford. He was in a hurry to work out the details of his idea. Arriving at his desk, he began the day by writing in his notepad the words: "Spectacular realization." Guth had intuited a mechanism that could explain how the Big Bang works.

Advanced concepts of general relativity and quantum mechanics are required to understand this "spectacular realization." Therefore, from this chapter onwards, our path will become more arduous, and a certain amount of concentration and patience will be necessary. However, it will be well worth the effort, because what is at stake is to actually witness the Big Bang in action, to understand the mechanisms that generated it, and to contemplate the universe before the advent of the Big Bang.

The Stuff of Angels

To understand the inner workings of the Big Bang, we need to know about a strange physical phenomenon that concerns the world of quantum mechanics. Certain types of particles, under special conditions, have the unique property of spontaneously aggregating into a state that physicists call a *field condensate with nonzero vacuum expectation value*. For the present purposes, this name is somewhat long-winded and not very suggestive. Therefore, I will refer to it here as a *vacuum substance*, to indicate its real essence and, at the same time, the fact that it exists in empty space.

To explain what the vacuum substance is, we must understand what is meant by the term 'elementary particle.' In the quantum description of the microscopic world, particles are interpreted as lumps of energy, called *quanta*, in an entity distributed throughout space–time, called a *quantum field*.

The concept can be made a little more familiar if we consider electric or magnetic fields, well known to anyone who did some physics experiments at school. Electromagnetic fields are continuous media that extend through space and propagate over time. However, when they are examined through the lens of quantum mechanics, as appropriate to the microscopic world, it turns out that they have a corpuscular nature. So, at short distances, the electromagnetic field has a structure made up of particles, called photons— the quanta of the electromagnetic field.

The vacuum substance is a quantum field composed of particles, just like the electromagnetic field. However, there is an important difference. An electric or magnetic field seen from different points of view—for example by observers in relative motion—changes in appearance. In fact, a static charge looks like an electric current to a moving observer, and therefore the electric and magnetic fields can swap roles if measured by different observers. This is perfectly natural: the presence of any real object in space is revealed precisely by observing it from different points of view. Surprisingly, this is no longer true for the vacuum substance, which looks exactly the same however we choose to observe it. Whatever the speed of the observer in uniform motion, or however our view of space is rotated or translated, the vacuum substance always looks the same. In short, the vacuum substance is indistinguishable from an empty space–time.

This is its extraordinary feature. The vacuum substance blends into empty space–time, as though camouflaged, becoming part of its structure. It is a collective phenomenon in which particles are arranged over an extended region of space. Although it is composed of particles, and therefore of matter, the vacuum substance is not a form of matter immersed in space but constitutes part of the very fabric of space–time.

The aggregate of particles comprising the vacuum substance has a physical reality and contains energy. Since this energy pervades space–time uniformly and persists even in the absence of any form of matter or radiation, it is known in physics as the *vacuum energy*. The name is well chosen because this surprising form of energy is truly woven into the texture of empty space–time.

The terms 'vacuum substance' and 'vacuum energy' are often used almost synonymously, as they are different facets of the same physical phenomenon. However, it is better to distinguish them: the former refers to the structure that fills space–time, while the latter describes its energy content. Metaphorically speaking, the vacuum substance is like the engine of a car, while the vacuum energy corresponds to the thrust that makes the car move.

The vacuum substance corresponds to a uniform but nonzero quantum field spread throughout space–time. The uniformity of the field ensures that space–time has the same characteristics as the vacuum, but the fact that it is nonzero indicates that there are particles there, containing vacuum energy. In other words, the vacuum substance is a field that uniformly pervades the whole of space–time, making it look empty, but concealing a real physical entity.

All of this may seem rather abstract. But maybe metaphysics can help us to build a mental picture. Let's try with a few examples.

In the eighth century, the Christian Arab theologian John of Damascus stated that the substance of which angels are made is incorporeal and immaterial compared to men, but is tangible and material in relation to God, because only the divine is truly incorporeal and immaterial. In some sense then, the vacuum substance is rather like what angels are made of, because it has a material reality compared to nothing, but it is the void in comparison to matter.

The Sanskrit word *sunyata* indicates emptiness. In Buddhist philosophy, *sunyata* is not the same as absolute nothingness, nor the negation of existence; it is the essential nature of things. It is also a meditative state, in which one achieves detachment from the impermanence of reality and an emptying of the mind that leads to wisdom. The many facets of *sunyata* are expressed in the verses of the *Heart Sutra*: "Form is emptiness, emptiness is form." *Sunyata* has subtle similarities with the vacuum substance which, while embodying the essence of empty space–time, is not nothing and has a real physical content expressed by the energy of the vacuum.

Perhaps it is better to leave metaphysics aside and return to physics. Although it seems rather abstract, the vacuum substance is in fact very real. One of the most concrete examples is superconductivity, that is, the property of some materials to conduct electric currents without any resistance when cooled to temperatures a few degrees above absolute zero. In the case of superconductivity, the vacuum substance is constituted by a particular arrangement of electrons that forms spontaneously within the material when the temperature drops below a critical value.

If one day we find a way to manufacture materials that have superconducting properties even at room temperature, it will bring about a genuine technological revolution that could transform our everyday lives. Electrical energy could be stored and distributed over long distances without loss. We would see nuclear fusion power plants proliferate, and we could wave goodbye to fossil fuels. Railway stations would be teeming with superfast magnetic levitation trains. Computers would reach unprecedented calculation speeds and electric motors would become superefficient. But even without dreaming of such a prosperous future, low-temperature superconductivity is already a reality, used for magnetic resonance scans in hospitals and in particle accelerators.

Another spectacular proof of the existence in nature of the vacuum substance phenomenon was found in 2012 with the discovery at CERN of the Higgs boson. It was thus shown that space–time is permeated with a vacuum substance made up of Higgs bosons, generated in the universe a tenth of a billionth of a second after the Big Bang.

The formation of vacuum substances seems to be a general phenomenon in particle physics, not limited to the Higgs boson alone. Therefore, it is not so hard to imagine that there were other forms of vacuum substance in the primordial universe, perhaps made up of some as-yet unknown species of particles.

The Anti-gravity of Vacuum Energy

The vacuum energy contained in the vacuum substance has a physical property that makes it absolutely unique compared to matter, radiation, or any other known form of energy. This exceptional property concerns gravity.

According to Einstein's relativity, gravity acts not only on the mass of a body, as is also predicted by Newtonian theory, but on any form of energy, including light. The observation of the gravitational deflection of light rays during the solar eclipse of 1919 was precisely what gave the decisive confirmation of general relativity.

The differences between Einstein's and Newton's theories do not end with the gravitational effect on all forms of energy. A further novelty of general relativity is that pressure also exerts a gravitational force. It is well known that a pressure difference exerts a force, clearly demonstrated every time we uncork a bottle of prosecco. But this is not the force described by Einstein's equation. According to general relativity, pressure causes a gravitational force, just like mass or energy. This has a truly striking consequence: vacuum energy produces anti-gravity!

To understand this, consider first a gas in a container, as shown in Fig. 9.1a. The gas exerts a positive pressure on the walls of the container, that is, a force that opposes any attempt to reduce the volume of the gas. According to general relativity, the pressure of the gas produces an attractive gravitational force in addition to that due to the mass and energy of the gas. The overall

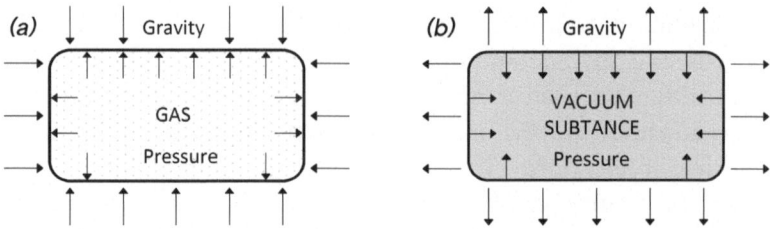

Fig. 9.1 **a** A gas exerts a (positive) outward pressure and gravity tends to make it implode. **b** A region filled with vacuum substance exerts a (negative) inward pressure and gravity tends to make it expand

gravitational force attracts the gas molecules toward each other, favoring their collapse, just as happens in star formation. So far, there is nothing untoward here.

But something strange happens when we consider a region of space permeated by a vacuum substance (see Fig. 9.1b). Vacuum energy behaves like an elastic material: a force must be applied to extend it. For this reason, vacuum energy exerts a force that opposes any attempt to expand the volume of space, that is, it produces negative pressure. Applying Einstein's equation, we find that the gravitational force due to this negative pressure of the vacuum substance is three times greater than the gravitational force due to its energy, and in the opposite direction. The net result is a repulsive gravitational force, in other words, the opposite of the familiar attraction due to gravity.

This result is worthy of a science fiction film. We are used to seeing things fall to the ground when they slip from our hands. We are used to thinking that planets must remain forever trapped in the Solar System. Our intuition tells us that everything must work this way and that gravity must always be attractive.

But no. Vacuum energy provides a surprising exception. A region of space pervaded by a vacuum substance is subject to gravity in reverse, or anti-gravity, which sends everything flying away in a dizzying expansion, driven by a repulsive force that actually grows with increasing distance. It's as though there's an invisible wind pushing every point in space further and further away from every other point in an overwhelming crescendo.

In 1882, the Irish writer Oscar Wilde visited the Niagara Falls. However, he was not impressed. With reference to the site's popularity as a honeymoon destination in those days, he defined it, with his unmistakable sarcasm, as "the second greatest disappointment to every American bride." Thoroughly bored by the visit, he remarked that it was just a huge amount of water flowing pointlessly from top to bottom, adding ironically that he would have been much more impressed if he had seen the water flowing the other way. At least on this last point, a good dose of vacuum energy could have mitigated Wilde's disappointment, because anti-gravity really does act in the opposite direction to gravity.

The attentive reader will certainly have noticed the deep similarity between vacuum energy and the cosmological constant, introduced by Einstein in his clumsy attempt to halt the collapse of the universe. Such an astute reader would be absolutely right, because this is exactly the same phenomenon.

Although it was his idea, Einstein was initially unhappy with the cosmological constant, and then subsequently even opposed to it, because he felt it was just a mathematical device, with no real physical meaning. But our understanding of the microscopic world has changed all this. When particle physicists discovered the vacuum energy, they identified the real meaning of the cosmological constant. It is not just a mathematical device, but a very precise physical concept, viz., the energy of empty space. It is a real entity, generated by a condensate of particles, and it really does produce the unearthly effect of anti-gravity, powering an unstoppable expansion of space.

There is a subtle difference though. The cosmological constant is, as its name suggests, constant: it does not change over time and it is the same everywhere in space. The vacuum energy, on the other hand, due to the dynamic structure of the vacuum substance, can vary over time and differ in different regions of space during the evolution of the universe. In this sense, vacuum energy generalizes the concept of the cosmological constant.

Spectacular Realization

The young Arthur, a simple squire, casually pulled a sword out of a rock, accomplishing a feat that had not been achieved by any of the most valiant British knights. He was thus recognized as the true king, the sole heir of Uther Pendragon, and found himself the proud owner of a magical sword. In the same way, Guth, casually reflecting on the cosmological effects of the vacuum energy, found himself in possession of a prodigious weapon. Now it was just a matter of working out what to do with it.

His "spectacular realization" was to imagine that, before the Big Bang, the universe might have been completely empty, dark, and cold: a true vacuum, without any form of matter or radiation. There was nothing, except for at least one speck of space containing vacuum energy. The space within that speck would have started expanding vertiginously, driven by anti-gravitational repulsion. With each regular beat of time, the distances between any two points would double. An expansion of this kind, doubling at regular intervals, is said to be *exponential*.

Exponential growth is often encountered in statistics and the term has now become familiar in everyday language. But despite this familiarity, it is not always appreciated just what is involved in exponential growth. Some examples will help to illustrate this.

Tear a page from this book and fold it in half, thus doubling the thickness. Now, fold this in half again. The thickness doubles once more, and there are four layers of paper. Fold them in half again. The thickness doubles, and there are eight layers of paper. Go on like this, repeating the operation forty-two times. How thick is the stack of paper at this point? The answer is that the stack is now high enough to reach the Moon! Try it if you don't believe me. Only forty-two folds are needed to obtain a stack of astronomical height. This is what is so amazing about exponential growth, in which the thickness doubles at every step. If this is not enough to convince you, let me try with another example.

A financial advisor offers you an investment with exponential growth: your capital will double every month. It seems like a good deal and you invest all your savings, which amount to a full thirty cents. It is indeed an exceptionally good deal and, provided that the advisor was not cheating you in some way, your capital will equal the entire world's gross domestic product after only four years!

Assuming that the reader is now convinced, let's go back to the expansion of space fueled by vacuum energy, which is equally prodigious. Using plausible values for the doubling time, it turns out that, during the phase preceding the Big Bang and in a time of only a quadrillionth of a nanosecond, a portion of space no larger than an atom could have transformed into a region as large as the Virgo supercluster, the gigantic cluster of galaxies in which we live. Guth called this crazy expansion process *inflation.*

Guth was inspired by the economic term used to indicate a general increase in prices, a topical issue at the time the theory was proposed. In 1979, inflation in the United States had reached 11%, but there are much worse cases in history. During the economic crisis in the Roman Empire in the third century, some goods became thousands of times more expensive within just a few decades. Inflation in the Weimar Republic was so rampant that, in 1923, the exchange rate reached four thousand billion German marks for one American dollar. We are used to thinking that the consequences of inflation are generally harmful: erosion of savings, decrease in purchasing power, slowdown in investments, and increase in public debt. For cosmic inflation, it's a whole different story and its consequences are astonishing.

The cosmic expansion predicted by inflation is so fast that the size of the initial speck of space containing vacuum energy makes little difference. It could have been microscopic or it could have filled the whole universe without making much difference to cosmic history. It only matters that at the beginning there was a place somewhere in the universe permeated by vacuum energy. Inflation then takes care of making that region grow out of all proportion, until it ends up dominating every other region of space.

If the universe before the Big Bang contained nothing but vacuum energy, we can easily deduce its geometry. As we saw in Chap. 2, de Sitter derived the shape of his space–time by assuming that the universe contained no matter of any kind, only a cosmological constant. Since the vacuum energy gives physical meaning to the cosmological constant, we arrive at the surprising conclusion that, before the Big Bang, the universe must have had the de Sitter geometry.

It is curious how the de Sitter space, invented in a misguided attempt to make the universe static and then abandoned in a dusty corner of the history of science, has reappeared many years later in a completely different guise. Who knows what de Sitter would make of it if we told him that the universe before the Big Bang was just like the space he had imagined.

For clarity, let me summarize what has been said so far. According to the theory of inflation, the universe before the Big Bang was just nothingness, except for vacuum energy wrapped into the very fabric of space–time. The immaterial and ineffable vacuum substance that permeated the universe powered an exorbitant dilation of space which inexorably diluted everything that was not vacuum energy. The primordial universe was a bleak nothingness of frenetically expanding empty space.

This portrait of the universe before the Big Bang—as disturbing as the work of some brooding existentialist poet—raises an immediate scientific question. How can the vacuum energy keep reproducing itself in such a way as to maintain the expansion of the universe in perpetual exponential growth? How can inflation sustain itself without exhausting its fuel?

An Inexhaustible Source of Energy?

A gas contained in an expanding vessel becomes progressively more and more rarefied. This is because the volume increases, while the amount of matter inside remains constant, whence the matter density decreases as the expansion proceeds. So far, everything is perfectly logical.

But something quite counterintuitive happens with vacuum energy. When a region of space filled with the vacuum substance expands, driven by its anti-gravity, the density of the vacuum energy is not reduced at all, but remains exactly the same. How is this possible? In order for the energy density to remain constant in an ever larger volume, the total amount of energy must also keep increasing. Think of a city where the outskirts are gradually extended with new constructions in order to maintain a constant population density (so, the number of inhabitants per square kilometer is kept the same). During this expansion, the population of the city must also increase.

The perpetual expansion of space while maintaining a constant vacuum energy density leads to an apparent paradox. What provides energy to the vacuum so that it can keep replicating during the exponential growth of space? Have we discovered an inexhaustible energy source? Perhaps energy was not conserved before the Big Bang?

The solution to the paradox lies in the fact that gravity hides within itself a form of energy. An analogy will help us to understand what is happening. Imagine Galileo bounding up the tower of Pisa at great speed with a cannonball under his arm. He then drops his load from the top. At the moment it slips from his hands, the cannonball is stationary and therefore has zero kinetic energy. As it falls, it gains speed and its kinetic energy increases. Has Galileo perhaps created energy from nothing? This is clearly not the case. To restore the energy balance and check that energy is conserved, we must take into account gravitational energy. When he climbed the spiral staircase of the leaning tower, Galileo gave the cannonball gravitational energy. After all, it must have been quite hard work carrying it to the top. As the cannonball fell, the gravitational energy was transformed into kinetic energy.

It is natural to assign the cannonball zero gravitational energy when it is infinitely far from the Earth, because at infinite distance the ball no longer feels any gravitational attraction. Having adopted this convention, the gravitational energy of the cannonball is always negative. From whatever point it falls, the cannonball acquires a positive kinetic energy that increases over time, at the expense of the gravitational energy, which becomes increasingly negative.

In the case of inflation, the secret lies in the anti-gravity exerted by the vacuum substance. Unlike normal gravity, whose energy becomes more negative when bodies approach, the gravitational energy of the vacuum substance becomes increasingly negative as space grows. Apart from this difference, the situation is analogous to the one described above. When the cannonball falls, its kinetic energy increases at the expense of its gravitational energy, which becomes increasingly negative. In the expansion of space, the vacuum energy increases at the expense of gravitational energy, which becomes increasingly negative.

During inflation, gravitational energy is converted into vacuum energy, compensating for the growth of space and maintaining a constant energy density of the vacuum substance. It is a perfect balance that allows the system to keep going, fueling an exponential expansion of space that leaves the energy density of the vacuum unchanged. Silently, gravity provides the energy to keep this seemingly miraculous process alive, without violating any physical law.

A mere speck of space containing vacuum energy is enough to trigger the chain reaction of inflation. The anti-gravity of the vacuum substance expands the space within the speck. In turn, the conversion of gravitational energy into vacuum energy replenishes the space that has just been created, keeping the vacuum energy density constant. And so it goes on, in an unstoppable whirlwind, regularly doubling all distances and transforming the initial speck into a volume larger than our entire universe, perhaps even larger than anything we could ever imagine. Essentially, inflation is a transfer of energy from gravity to empty space.

The End of Inflation

At first glance, what inflation tells us about the universe before the Big Bang seems to have little to do with the real world. However, as we shall see later, the frenzied expansion of that cold, dark, empty space was quietly preparing the perfect conditions for a universe as diverse and complex as the one we live in. Then, we are compelled to ask: what transformed the solitude of empty space *before* the Big Bang into the seething soup of matter that would eventually spawn the stars and planets *after* the Big Bang?

Initially, Guth imagined a situation rather like what happens in the transition between water and ice, known as a phase transition. The same substance can exist in a liquid phase (water) or a solid phase (ice), depending on the temperature and pressure. Similarly, space–time can exist in different states, depending on external circumstances: in one state, elementary particles form a vacuum substance, while in another, there is nothing and the vacuum energy is zero.

The expedient Guth proposed to end the inflation process was based on fortuitous transitions between different states of space–time. The vacuum energy would suddenly disappear, interrupting the chain reaction that fuels inflation. This process may sound like some form of magic, but it is a perfectly well-known phenomenon in quantum mechanics, according to which a physical state can undergo a sudden transformation into a different state. Such a phenomenon occurs in the decays of certain radioactive nuclei, and also in electronic components such as tunnel diodes or Josephson junctions.

Unfortunately, the mechanism devised by Guth to stop inflation does not work, because it turns out that it could not completely bring it to an end. Somewhere in the universe, inflation would always stubbornly continue, a bit like a fire that can never be completely put out because, extinguished in one place, it starts up again in another.

The problem was solved in 1981 by a young Russian cosmologist, Andrei Linde, and a similar solution was proposed independently a few months later by Andreas Albrecht and Paul Steinhardt of the University of Pennsylvania. The theory of inflation is today one of the most active fields in cosmology, and hundreds of researchers have made significant contributions to it. Among these, Linde stands out as one of the most iconic figures for his many fundamental contributions to the development of the theory, from the beginning right up to the present. Charismatic, eclectic, and ingenious, Linde worked at CERN in the late 1980s, before moving to Stanford University, where he remains to this day. He has a sharp wit which he uses generously in scientific discussions. With his deadpan expression and pronounced Russian accent, he alternates between hilarious jokes and profound scientific insights. He is an endless source of ideas and many of the advances in the field of inflation are down to him.

Linde and his colleagues started from a simple observation. There has to be a precise relationship between the strength of the vacuum substance and the vacuum energy density. It's like a radiator: the position of the knob (the strength of the vacuum substance) determines the temperature of the radiator (the vacuum energy density). In physics jargon, the mathematical relationship that links the vacuum energy to the vacuum substance is called a *potential*. The form of this mathematical relationship depends on the microscopic structure of the particle condensate constituting the vacuum substance. The potential is like the instruction manual of the radiator, which tells us what the temperature will be, given the position of the knob.

During cosmic evolution, the strength of the vacuum substance does not remain constant but varies according to a physical law that depends on the potential. It's like having a faulty knob which, however it is set, always gradually turns of its own accord until it reaches the same final position.

Today, it is thought that cosmic inflation ended through a variant of Guth's initial idea. Instead of having an abrupt transition between two distinct states, the vacuum substance is thought to have varied very slowly throughout the history of the universe, until it finally disappeared. This gradual metamorphosis would allow the inflationary process to come to a halt everywhere in space, without running up against the difficulties that Guth encountered with his instantaneous phase transition.

An analogy can help us to picture this. Imagine a marble moving at the bottom of a slightly concave bowl, whose surface exerts a strong friction (see Fig. 9.2). In this analogy, the shape of the bowl corresponds to the potential, that is, the relationship between the strength of the vacuum substance and the vacuum energy. The position of the marble represents the state of the universe. When the marble is at the bottom of the bowl, there is no vacuum substance and the vacuum energy density is zero, so inflation ceases. The further the marble is from the bottom of the bowl, the greater the vacuum energy, making the inflationary process faster. The friction acting on the marble corresponds to the effect of the expansion of space, which slows down the evolution of the vacuum substance.

The analogy fits perfectly, because the equation that describes the evolution of the vacuum substance during inflation has the same mathematical form as the one describing the motion of the marble, despite the physical phenomena being completely different.

To know the shape of the potential, we would need to know what kind of particles constitute the vacuum substance that causes inflation, but unfortunately, we don't. In the language of the previous metaphor, nature has forgotten to tell us where it left the instruction manual for the radiator and we have not yet found it. On the positive side, many of the consequences of inflation can be deduced by making simple assumptions about the potential, without knowing its exact expression.

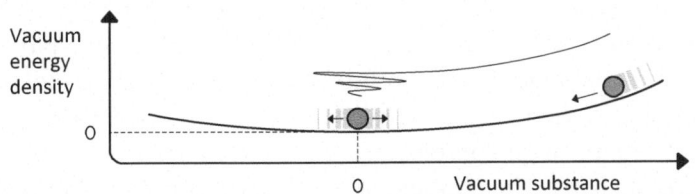

Fig. 9.2 The motion of a marble in a bowl with friction is used as an analogy for the cosmic evolution of the vacuum substance. The shape of the bowl represents the potential, that is, the relationship between the vacuum substance and the energy of the vacuum. The position of the marble along the horizontal axis represents the strength of the vacuum substance. The bottom of the bowl corresponds to the state in which the vacuum energy density is zero, and the height relative to this point defines the value of the vacuum energy density of the corresponding state

To give a concrete example, imagine that the potential has the form shown in Fig. 9.2. It is not difficult to see how the marble will move. If it is placed near the edge of the bowl, it will roll slowly toward the bottom where, after a few oscillations up and down the slope, during which it is gradually slowed down by friction, it will eventually come to a halt. The same thing would have happened in the universe before the Big Bang. Starting from an initial value that triggered inflation, the energy density of the vacuum slowly decreased until it could no longer sustain the exponential expansion process.

During the last few oscillations around the final state, the particles frozen inside the vacuum substance would come to life, converting themselves into other kinds of particles. These quantum transformation processes may seem mysterious, but they are observed all the time in experiments with large accelerators. Their study is one of my favorite pastimes, and it is what happily fills the days of my physicist colleagues at CERN.

At the moment when inflation switches off, the vacuum substance transforms into a flurry of particles and the vacuum energy is released as the thermal energy of the incipient primordial gas. From the ashes of the vacuum substance, matter is born. At this very instant, the cold and empty universe transforms into a boiling mass, thronging with all the constituent elements of matter and radiation that populate the cosmos today. This was the instant of the Big Bang.

The Big Bang According to Inflation

As explained by inflation, the inner mechanism of the Big Bang is such an extraordinary story that it deserves a recap. According to this theory, the Big Bang is the moment when the energy stored in the fabric of empty space–time undergoes a quantum transformation, which breathes life into the fundamental components of matter. It is the moment when the cold emptiness of space, at a temperature of absolute zero, is suddenly populated by an extremely hot gas, rich in all the basic elements needed to generate stars, planets, and life, through complex processes lasting tens of billions of years.

The Big Bang is also the moment when the astonishing expansion of space, doubling in size at every beat of time, finally comes to an end. From that time on, the universe continues to expand due to the inertia of the initial push received during the inflationary phase, just as a toy car will continue to roll across the floor even though it is no longer being pushed by a child's hand.

Ironically, according to inflation theory, the Big Bang was not the explosive beginning of an expansion phase in the universe. On the contrary, the Big

Bang marked the end of a frantic exponential expansion. After this event, the universe maintained only a faint echo of its dramatic inflationary existence. Space went on expanding at an ever more moderate rate, steadily slowing down due to the gravitational attraction exerted by matter.

The theory of inflation clarifies the point that the Big Bang was not an explosion that occurred at a point in space, but rather a uniform transition, which involved almost simultaneously the entire region of space that today forms our observable universe, and probably much more. This transition corresponds to a transformation of space–time from a state permeated with vacuum energy to a state completely devoid of it; from a state in which particles were trapped in a uniform condensate, to one in which particles move frantically to and fro in thermal agitation.

The Big Bang that emerged from inflation was nothing like a gigantic explosion. The picture we have is rather that of an immense frozen lake that suddenly melted, changing state everywhere in space; or a colossal house of cards that collapsed under a violent gust of wind; or again, a mountain that subsided in a colossal landslide, leaving behind the debris that would give rise to a new form of physical reality.

The portrait of the Big Bang painted by inflation theory may seem vague and abstract, a fantastic story flirting with metaphysics. But, to understand whether inflation deserves the title of scientific theory, we must put it to the test. It is not enough for a scientific theory to be able to tell a story. It must explain experimental data, provide a deeper understanding of natural phenomena than previous theories, and offer solutions to unresolved enigmas. The time has therefore come to see whether inflation passes these tests and ask whether it deserves to be qualified as a scientific explanation of the Big Bang.

10

Unveiling the Mysteries of the Big Bang

> *Can we actually 'know' the universe? My God, it's*
> *hard enough finding your way around in Chinatown.*
> Woody Allen

As a boy, I was thrilled by the stories of the archaeologist Heinrich Schliemann. I was fascinated by his obsession to prove that the *Iliad* is not just the fruit of Homer's imagination, but tells of events that really happened. Years ago, I stayed in Mycenae in the same room that Schliemann had occupied during the excavations in which he found the golden mask of Agamemnon. I felt the thrill of sleeping in the same wrought iron bed used by the eccentric German archaeologist, dreaming of the deeds of Achilles and Hector, as perhaps he had done himself a hundred and thirty years earlier.

Schliemann identified the site of the mythical Troy and brought to light not one city, but nine layers of cities buried one on top of the other. Today his excavation methods are heavily criticized and there are those who say with sarcasm that Schliemann achieved what the Achaeans had failed to do: destroy the walls of Troy. But my boyhood fascination for Schliemann came from the way he pursued a crazy idea and ended up discovering much more than even the mind of a visionary could have imagined. This is what happened with the theory of inflation.

The original version of the chapter has been revised. Missed text corrections and figure correction have been implemented. Further details can be found at
https://doi.org/10.1007/978-3-031-69933-7_18

At the time of its discovery, inflation seemed only a bold idea that could at best resolve certain abstract theoretical paradoxes. As the research continued, new levels of understanding emerged, along with some completely unexpected results. Today, we know that the idea of inflation resolves many of the mysteries raised by the Big Bang theory (see Chap. 8).

Solution to Mystery Number 1: The Expansion

Inflation offers a clear answer to the question of what triggered the expansion of space. The evolution of the universe was driven by a repulsive gravitational force produced by the vacuum energy. The condensate of a hypothetical elementary particle modified the structure of space–time before the Big Bang, providing the vacuum energy to power an unbridled expansion of space. This prodigious phenomenon ended with the Big Bang, while the expansion of the universe discovered by Hubble is only a faint reminder of events that took place in the remote past. The explanation for what powered the Big Bang is therefore concealed in the innermost confines of the world of subnuclear particles.

The Cosmic Horizon

Before addressing the other mysteries of the Big Bang, a digression is necessary to explain the idea of a cosmic horizon.

When we stand on the shore and admire the sea, we may feel a certain serenity and instinctively reflect on the meaning of the cosmos. Perhaps we entertain such deep thoughts unconsciously because the sea seems to have no bounds. In reality, from the shore, we cannot see further than five kilometers. Beyond this limit, the sea is invisible, hidden by the Earth's curvature.

Viewing from higher up, the horizon moves further away. Climbing up to the crow's nest of a ship, at a height of about forty meters, the sea becomes visible up to a distance of about twelve nautical miles, or twenty-two kilometers. It is no coincidence that international law has chosen precisely the distance of twelve nautical miles as the limit of territorial waters, because this is the maximum distance that can be monitored visually with a good pair of binoculars. Using more powerful optical instruments would serve no purpose. Beyond that distance, the sea disappears behind the horizon due to the Earth's curvature, and its surface remains irremediably invisible, as illustrated in Fig. 10.1a.

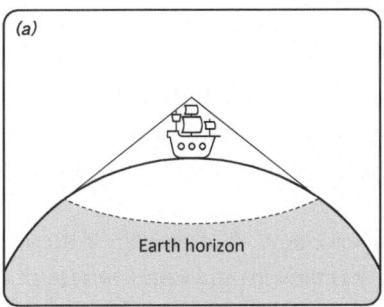
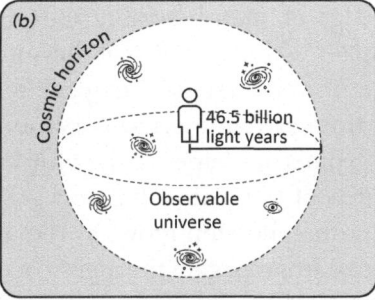

Fig. 10.1 a The Earth's horizon is a circle on the surface of the Earth with the observer at the center. **b** The cosmic horizon is a spherical surface in space with the observer at the center. The three-dimensional space contained within the cosmic horizon is called the observable universe

A similar phenomenon occurs in the universe, although it has nothing to do with the Earth's curvature. According to the theory of relativity, no physical information can propagate faster than light, that is, faster than one billion kilometers an hour. Since the time elapsed since the Big Bang is not infinite, there is a maximum distance—bounded by the *cosmic horizon*—from which we can receive signals. The universe beyond the cosmic horizon is irremediably invisible, because no signal would have had time to reach us, not even if it had started traveling towards us at the time of the Big Bang. For this reason, the region of space within the cosmic horizon is called the *observable universe*. Everything beyond the boundaries of the observable universe is invisible, not due to insufficient resolution of astronomical telescopes, but due to an absolute limitation imposed by physical laws. In a sense, the cosmic horizon marks the limit of territorial waters accessible to human observation.

If the universe were not expanding, the cosmic horizon would be 13.8 billion light-years away from us, because this is the distance that an electromagnetic signal travels in a time equal to the age of the universe from the Big Bang to today. In reality, the observable universe is larger. Due to the expansion of space, the source of a light signal is today further away from us than at the time it emitted the signal. Taking this effect into account, we find that the cosmic horizon is 46.5 billion light-years away from us.

The observable universe is therefore a sphere of radius 46.5 billion light-years, with the Earth exactly at the center (see Fig. 10.1b). Only a stubborn anthropocentric mindset could believe that the observable universe corresponds to the entire universe, with humanity sitting bang in the middle. The boundaries of the observable universe are clearly only a consequence of our particular point of observation. Just as the sea does not end at the limit of our territorial waters, there is no reason to believe that something special happens at the edge of the observable universe. Although invisible to our observations, space will continue well beyond the cosmic horizon. We can only say that the universe extends in space for at least 46.5 billion light-years from us, but from astronomical observation alone, we cannot know its real size or overall geometric shape. Only with logical deduction can we go beyond the boundaries of the cosmic horizon.

The cosmic horizon grows as the universe ages, because the time available for a signal to propagate through space increases. In universes where matter or radiation are the most common forms of energy, gravitational attraction slows the expansion of space, so, from a certain time onwards, the cosmic horizon grows faster than space expands. This is illustrated in Fig. 10.2, where the lines describe how the positions of equidistant points vary with time along one direction of space. As lines spread out, space is expanding because physical points move apart. As lines thicken, space is contracting because points come closer. The grey region indicates the observable universe, relative to an observer located at the centre, and its boundary is the cosmic horizon. The figure shows that, after the Big Bang, the cosmic horizon overtakes the expansion as time goes by, and the observable universe keeps on encompassing new regions of space. It's like looking at the sea while climbing up a mountain: the horizon will move further and further away and parts of the sea that were previously invisible will become observable. As we get to higher altitudes, ever more distant ships come into view.

It's the opposite when space expansion is accelerating, rather than decelerating. In this case, the cosmic horizon is determined by the distance at which the recession speed of space equals that of light. For an accelerated expansion, in contrast to the decelerating case, points in space that are flying away from us faster than light will disappear forever from our sight, as their speed continues to increase.

De Sitter's geometry, which describes the inflationary epoch, is such that the radius of the observable universe remains constant over time, while space is in continuous exponential growth, quickly escaping beyond the boundaries of the cosmic horizon. This is what happens during the epoch before the Big Bang, as shown in Fig. 10.2. It's like standing on the beach and looking out over the sea. The horizon always remains the same, but a strong current is dragging everything offshore, carrying the ships beyond the horizon. What was previously visible, disappears from our sight forever.

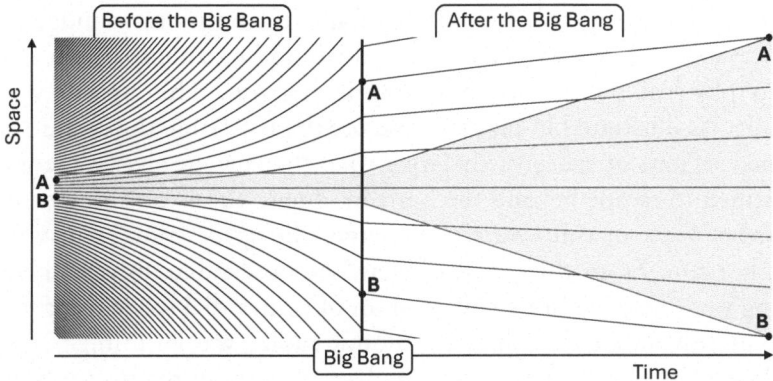

Fig. 10.2 A simplified sketch of the evolution of cosmic distances along one direction of space. The grey area denotes the observable universe, relative to an observer located at the centre. Before the Big Bang, space expands exponentially, escaping out of the cosmic horizon which remains constant. After the Big Bang, the horizon grows faster than space expansion and the observable universe keeps on incorporating new spatial regions. Points A and B, which lie on our current horizon in opposite directions in the sky, have been outside the horizon throughout cosmic history since the Big Bang, but were within the horizon in the distant past, sufficiently before the Big Bang

Did the Universe Start Out Small?

As we go back in time towards the Big Bang, the distances between physical points contract and the radius of the observable universe decreases. Since we still do not know the value of the vacuum energy during inflation, we cannot calculate the size of the observable universe at the time of the Big Bang, but we know that it could have been billions and billions of times smaller than an atom. For this reason, we often read in the newspapers that, at its origin, the universe was no more than a tiny dot. But is this really so?

The observable universe is the region of space in causal contact with us, that is, within which physical information can be exchanged. It tells us how far communication can propagate, but nothing about the actual size of the universe. Therefore, for all we know, at the moment of the Big Bang, the universe as a whole could have been huge, even infinite.

Moreover, it means nothing to say that the observable universe is large or small, unless we specify a unit of measure for comparison. We instinctively decide whether a visible object is large or small by comparing it with the human dimension, and when dealing with microscopic phenomena, we might compare with the size of an atom. However, at the time of the Big Bang, the universe contained neither humans nor atoms, so the comparison is meaningless.

Since space was empty before the Big Bang, the only way to judge whether the observable universe was large or small would be to compare the distances between physical points with the cosmic horizon. The answer we get is surprising. As illustrated in Fig. 10.2, before the Big Bang, the cosmic horizon contained regions of space vastly larger than the current observable universe, but destined to escape beyond the horizon, due to the astonishing expansion. The further back in time we go, the larger the space contained within the cosmic horizon. From this point of view, the observable universe before the Big Bang was anything but small. A grain of primordial space measuring just a few billionths of a billionth of the atomic radius was an immense expanse, much larger than all the space we can observe today in the universe. Perhaps the visionary poet William Blake had just this intuition when, without knowing anything about the theory of cosmic inflation, he spoke of seeing "a world in a grain of sand."

Solution to Mystery Number 2: The Uniformity

This mystery concerns the paradoxical observation that the cosmic background radiation is uniform across the celestial vault, even in regions that have not had time to communicate with each other. For example, consider two points A and B, opposite each other in the sky, which emitted cosmic background radiation that is reaching us just now (see Fig. 10.2). Today's distance between points A and B is equal to the diameter of our observable universe, or 93 billion light-years. Going back in time, the points A and B approach one another due to the contraction of space (the expansion seen in reverse), but the horizon contracts much more quickly, as shown in Fig. 10.2. The result is that, at the time when the cosmic radiation was emitted, points A and B were too far apart to have ever communicated with each other. It seems like a pure coincidence then that the temperature of the background radiation should be almost perfectly identical at A and B.

This paradoxical conclusion, which we called the mystery of uniformity, results from a hidden assumption. We tacitly assumed that the expansion of the universe was decelerating. This is a perfectly reasonable assumption for a universe composed of matter or radiation, because gravitational attraction can only slow down the expansion. But the anti-gravity effect produced by vacuum energy can do something utterly counterintuitive.

This is the crucial difference which provides the solution to the mystery of uniformity. In a universe with decelerating expansion, what was beyond the cosmic horizon in the past may one day come within it and thus enter the observable universe. In a universe with accelerating expansion, what was part of the observable universe in the past will disappear beyond the cosmic horizon in the future.

Even points in space that appear very widely separated today resided within the cosmic horizon before the Big Bang (see Fig. 10.2). A region much larger than the current observable universe was squeezed inside the same horizon and had all the time it needed to reach uniformity, just as the temperature can become uniform inside a closed room.

In Chap. 8, I compared the mystery of uniformity to the situation where a cup of coffee forgotten in the kitchen is found to have exactly the same temperature as all the cups of coffee left by aliens in their kitchens in every corner of the universe. Inflation lifts the veil on this mystery by explaining that those cups of coffee, now so far apart, were once all crammed inside the same cupboard, where they reached and have since maintained the same temperature.

In conclusion, according to inflation, the mystery of uniformity is just a misunderstanding, born from the prejudice that gravity must always be attractive. This prejudice leads us to believe that the current universe is a mosaic of regions of space that were completely disconnected in the past. On the contrary, before the Big Bang, a space enormously larger than everything we observe today in the universe was in close causal contact, thus allowing cosmic uniformity.

Solution to Mystery Number 3: The Flatness

The flatness conundrum concerns the instability of the flat geometry during cosmic evolution. However the universe sets out at the instant of the Big Bang, its geometry will quickly deviate from flatness and the curvature will become increasingly pronounced as the universe ages. To explain how the universe can be as flat as we observe it today, its flatness must have been almost perfect at the time of the Big Bang. The mystery here is to understand what mechanism could have flattened the universe to such an extreme degree.

As in the previous cases, it is once again the marvel of accelerated expansion that resolves the mystery. Rather than being unstable, as in the case of a decelerating expansion, the flat geometry is the inevitable result of an accelerating cosmic expansion. Whatever the initial geometry of space, after a period of inflation, the universe will be almost perfectly flat.

It is not hard to see why inflation would flatten the geometry. Consider for example the spherical surface shown in Fig. 10.3. During inflation, the radius

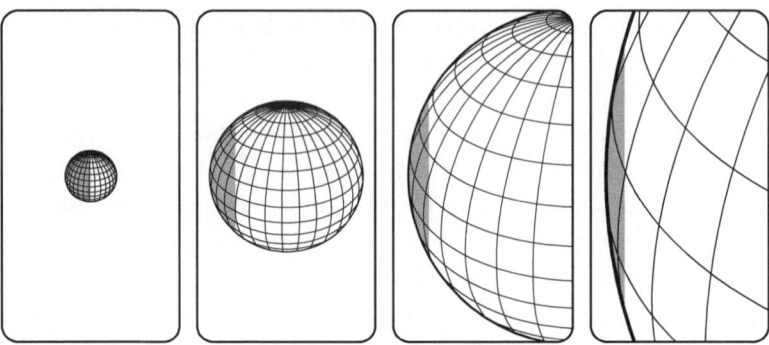

Fig. 10.3 The gray regions have the same area but are drawn on larger and larger spherical surfaces. As the radius of the sphere increases, the geometry of the gray region becomes flatter

of the sphere grows enormously while, as mentioned previously, the horizon remains constant. It is clear from the figure how the curvature of space within the horizon will quickly disappear over time, leaving an increasingly flat surface.

This works not only for a spherical geometry, but for any kind of geometry. Inflation acts on space as if it were pulling a stretchy fabric. Whatever its initial shape, the fabric will end up almost perfectly smooth if it is stretched for long enough.

The flatness conundrum is thus solved by inflation. We would be incredibly unlikely to observe a flat universe today if space had only ever undergone a decelerating evolution. Instead, an extraordinarily flat geometry is not only likely, it is even the only possible outcome of a sufficiently long period of inflation.

The Energy of the Universe

Einstein's equation tells us that the geometry of space is an expression of its energy content. Since inflation predicts a flat geometry, it issues a precise verdict on the density of energy in the universe. According to inflation, the current universe contains an energy equivalent to that of five hydrogen atoms in every cubic meter of space, or the mass of a fly in the volume of space occupied by the Earth. That's not much. After 13.8 billion years of steady expansion following the Big Bang, the universe has become so diluted that it finds itself almost empty today. Life on Earth is a sporadic oasis of matter immersed in an enormous desert. If all the energy of the universe were in the form of human beings distributed uniformly in space, the distance between each individual and their nearest neighbor would be about a million kilometers. Our social life would be terribly boring.

The extremely low energy density of the universe may provide the answer to a famous question by Enrico Fermi. During a lunch among physicists at the Los Alamos laboratory canteen in the summer of 1950, after a brief contemplative silence, Fermi blurted out: "But where is everybody?"

He was referring to the fact that, according to his own estimate, life should not be such a rare phenomenon in the universe. He therefore wondered why aliens had not already made themselves known, for example, by visiting the Earth. If I had been at the table with Fermi, I would have replied that, according to inflation, matter in the universe is so rarefied that, even if life were indeed a relatively common phenomenon, the probability of aliens living near the Earth is extremely low. It's a pity that the theory of inflation was not yet known in 1950.

We should note that, when Guth proposed the idea of inflation, measurements of the energy density of the cosmos suggested a value just over a quarter of what was predicted by inflation. Guth was not discouraged. The theory was too exciting to give in to a mere experimental observation. Perhaps he had learned from an anecdote according to which, many years before, Eddington is reported to have said: "Never believe an experiment until it has been confirmed by a theory." Eddington was certainly right as regards the cosmic energy density. Today we know that those early measurements were wrong, because they did not include the contribution of dark energy which, after all, had not yet been discovered. Current measurements of the average energy density of the universe are in spectacular agreement with the prediction of inflation. Guth had hit the nail on the head.

The Problem of Mystery Number 4: The Arrow of Time

In order for there to be a flow of time that allows the development of the complex evolutionary processes responsible for life, the universe must have been born in an exceptionally ordered state. This is, in essence, the problem of the arrow of time.

According to the inflation theory, the universe emerged from the Big Bang in a state of almost perfect uniformity, and hence of exceptional order (i.e., low entropy). In this way, inflation set the universe up with an arrow of time, allowing it to produce all those spectacular phenomena that we admire today, including spectators.

It may seem that, in creating a state of low entropy in the universe at the instant of the Big Bang, inflation must have violated the second law of thermodynamics, according to which entropy cannot decrease. But this is not the case. Despite its prodigious consequences, inflation did not violate any physical principle.

Calculations show that, taking into account the repulsive gravity and the enormous dilation of space, the entropy of the universe increased during inflation, fully complying with the second law. Nevertheless, the increase was so slow that at the instant of the Big Bang, when the energy of the vacuum was converted into thermal energy, the universe found itself in a state with much lower entropy than a generic universe made of matter and radiation, whence it appeared so ordered as to seem almost surreal.

But this is not enough to get to the root of the problem. For, if inflation explains the initial conditions of the universe at the time of the Big Bang, what explains the initial conditions of inflation? There is still no answer to this question and the theory of inflation only pushes this question about the origin of the arrow of time back to an earlier phase.

In conclusion, inflation cannot yet explain the mystery of the arrow of time. To solve the enigma—a fundamental question for our very existence in the universe—we would need to go even further back in the past and understand what happened before inflation.

11

Fossils of the Big Bang

Don't look for obscure formulas or mystery
in my work. It is pure joy that I offer you.
Constantin Brâncuşi

What if I were to tell you that all the structures observed today in the sky were born from seeds sown by quantum mechanics before the Big Bang? That those seeds remained frozen beyond the cosmic horizon, then came to life, billions of years after the Big Bang, to form stars, galaxies, and galactic clusters?

The story seems so improbable that we might think it was the work of a science fiction writer. But in reality, it stems from an advanced understanding of the physical laws that govern both the microscopic structure of space and the large-scale structure of the universe. It is the result of accurate mathematical calculations that extrapolate our understanding of reality back to the times of the Big Bang and beyond. That it is science, and not fantasy, is confirmed by observational data on the cosmic background radiation and the distribution of matter in the universe. And one day, perhaps, measurements of primordial gravitational waves will provide us with further confirmation.

This is one of the most fascinating chapters in the story of human scientific discovery. It is awe-inspiring to see how the world of elementary particles harmonizes with cosmic immensity, revealing the profound unity of the natural order. The story of how inflation explains the emergence of structure in the universe is one of those thrilling scientific results that can fill us with wonder, as we witness the spectacle of the birth of the universe. This is the story I will now recount.

© The Author(s), under exclusive license to Springer Nature
Switzerland AG 2024
G. F. Giudice, *Before the Big Bang*, Copernicus Books,
https://doi.org/10.1007/978-3-031-69933-7_11

Solution to Mystery Number 5: Cosmic Structure

Inflation is so successful in explaining some of the mysteries of the Big Bang because of the extraordinary way it stretches space, making everything uniform. At first glance, however, this suggests that it will encounter an inevitable failure, because it seems to preclude any explanation for the origin of cosmic structure. Indeed, if the universe were born from the Big Bang in a perfectly uniform state, it would remain so forever more, never able to generate any kind of celestial body. Gravitational collapse can amplify initial inhomogeneities, but it cannot create them from nothing.

The situation looks desperate, but here a completely unexpected *deus ex machina* comes to the rescue: quantum mechanics. This may seem surprising because quantum mechanics is the theory describing the microscopic world from elementary particles to atoms, so what could that have to do with the formation of galaxies, gigantic cosmic structures containing up to hundreds of millions of millions of stars? Given that quantum effects are practically invisible even on a human scale, it may be hard to imagine how they could be relevant to the structure of the universe as a whole.

But inflation is full of surprises. The frantic expansion of space completely transforms the notion of distance: what was microscopic at one moment becomes astronomically vast at the next, and this projects quantum mechanics into a leading role on the cosmic stage.

One of the cornerstones of quantum mechanics is Heisenberg's uncertainty principle, which expresses the probabilistic nature of the theory. According to this principle, the position and speed of a particle cannot be known simultaneously with absolute precision, not due to any inadequacy of our measuring instruments, but due to an indeterminacy inherent in the natural world. This is certainly a strange result, going against our normal intuitions, but quantum mechanics is full of surprising features. However, such oddities should not mislead us into thinking that quantum mechanics is some kind of fantasy. It describes the reality in which we live. Transistors, lasers, and microchips in computers are the living proof, because they all work thanks to quantum mechanics.

One consequence of Heisenberg's principle is that physical quantities such as the position and speed of particles are inevitably subject to *quantum fluctuations*. These fluctuations are random variations in those physical quantities, which make it impossible to predict with certainty the result of a measurement. The theory tells us only the probability of the outcome of a measurement. It's rather like playing a game of backgammon, where you can calculate the probability of taking one of your opponent's counters, but you can never be sure of succeeding until you have rolled the dice.

Quantum fluctuations contain information about the statistical uncertainty in the value of a physical quantity. They do not depend on the accuracy of our measuring instruments and cannot be eliminated by repeating the measurement, no matter how often. Quantum fluctuations reflect the enigmatic indeterminacy of objective reality in the microscopic world.

The impossibility of specifying with certainty position and speed in quantum mechanics rather undermines the concept of trajectory. In practice, it is as if it were impossible to focus on a moving particle. Its trajectory can only appear as a blurred image. With our senses, we perceive sharp images of reality because we observe things from afar. It is only when we look deep into the bizarre microscopic world that we begin to notice the blurring of trajectories due to quantum mechanics.

What applies to a particle, also applies to the vacuum substance, which is after all an aggregate of elementary particles. As a result, quantum mechanics imposes small but inevitable fluctuations on the vacuum substance. Its vacuum energy will not be perfectly uniform in space, but differ slightly from region to region. These regions are tiny, but the phantasmagoric expansion of space driven by anti-gravity—that is, inflation—amplifies them, transforming imperceptible specks into astronomical expanses which extend even beyond the observable universe and remain etched in the vastness of the cosmos, beyond the horizon.

After the Big Bang, the original vacuum energy fluctuations were transformed into tiny variations in the matter density, which extended over astronomically vast regions. At this point, the attractive effect of gravity would have come into play. Just as in the world the rich become richer and the poor become poorer, regions with a matter excess would have attracted more matter, at the expense of regions where matter was scarce. This gravitational collapse would have amplified the tiny variations in density, transforming them into the large-scale structures we observe today with our telescopes. The whole panoply of galaxies, clusters, and superclusters that shine in the sky today are the result of quantum effects that survived the Big Bang. This is how inflation solves the mystery of cosmic structure.

The complexity of the universe is generated by the physical laws of both the very small and the very large. Tiny variations in density predicted by quantum mechanics and a sufficiently long lapse of time since the Big Bang are all that is required for gravity to convert microscopic primordial seeds into complex worlds where life can arise. Inflation provides the link between the microscopic and cosmic realities. The structure of the universe arises from this interplay between the physical laws that govern the world of elementary particles and those that dictate the geometry of space. We really do live in a quantum universe.

If quantum mechanics did not exist, inflation would have generated a perfectly flat, uniform, and sterile universe like a vast desert. We owe our existence to the quantum seeds from which the complex structures of the current universe were born. The presence of these seeds is not an ad hoc adjustment of the theory of inflation, but an inevitable consequence of quantum mechanics. The fundamental laws of physics decree the success of inflation in solving the mystery of cosmic structure.

When you look up at the sky on a clear, moonless night, silently contemplating the beauty of the cosmos, are you moved? Are you impressed by the bold figure of Orion, with his sparkling belt and the sword hanging at his side? Are you pleased to find Cassiopeia up high in the firmament, the vain and boastful queen of Ethiopia who considered herself more beautiful than the Nereids, arousing the disastrous wrath of Poseidon? If your admiration of the sky stops here, you have missed the best part of the story.

Beyond the stars and much further away from us, there are galaxies. With the naked eye, it's hard to make out more than Andromeda, or the Magellanic Clouds if you live in the southern hemisphere. But scientific instruments have identified thousands of billions of galaxies and mapped their positions. What their distribution reveals are not the features of mythological heroes, but the pattern of quantum fluctuations in the vacuum energy that existed before the Big Bang. In this sense, they are like gigantic fossils from a primordial era.

This is one of the most astonishing stories ever told by science. Just as the fossils we find inside rocks preserve the trace of plants and animals that existed long before human beings populated the Earth, so the large-scale structures of the universe we observe in the night sky bear a record of the microscopic quantum fluctuations in the particle condensate that pervaded primordial space prior to the Big Bang. And just as we can reconstruct the shape of gigantic dinosaurs from scattered fossil remains, we can use astronomical measurements to reconstruct what was going on in the universe in times so remote that they seem way beyond the boundaries of human exploration. This is the amazing story told by inflation.

A Word of Warning

The above is just a summary of the main points put forward by inflation theory to solve the mystery of cosmic structure. However, the phenomenon is so fascinating that I think it deserves a deeper look. The rest of the chapter will be devoted to this. To do this, I will need to introduce some more technical concepts, so the following is aimed primarily at those who wish to know more about Big Bang physics and are ready to engage with more complex scientific arguments. Other readers can jump directly to the next chapter without losing the logical thread of the story.

The Fossils of Matter

The first ingredient in the inflationary recipe for cooking up cosmic structure is quantum mechanics, according to which the energy of the vacuum that pervaded the universe before the Big Bang cannot be perfectly uniform, but must inevitably be subject to random fluctuations. Such fluctuations are characterized by two quantities, as shown schematically in Fig. 11.1: the *amplitude* (i.e., the variation in the vacuum energy density) and the *extent* (i.e., the size of the corresponding region of space). When a fluctuation occurs, its extent is about equal to the cosmic horizon, because this is the maximum distance over which physical interactions can occur. The amplitude can be calculated in probabilistic terms from the laws of quantum mechanics. The miraculous feature of the inflationary expansion of space is that it catapults the spatial extent of a fluctuation to a colossal size, while leaving the amplitude unchanged, which means that the amplitude remains minuscule.

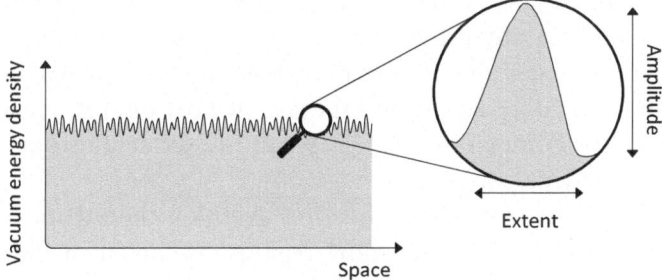

Fig. 11.1 Fluctuations in vacuum energy density can be viewed as modulations around a constant value. They are characterized by their amplitude and their spatial extent, as shown in the inset

Since quantum fluctuations were produced continuously during inflation, some would have had more time to expand than others. The result is that regions of every possible size were generated in the universe, with very slight differences in the vacuum energy density. After a long enough period of inflation, the universe would have looked like a map in a gigantic three-dimensional atlas, with countries of every possible size, from those as small as the Vatican City to those as large as Russia, while the vacuum energy density would have differed almost imperceptibly from one country to another.

Because of these small inhomogeneities, the Big Bang would not have occurred everywhere at the same time. In some regions, the vacuum substance would have switched off and triggered the Big Bang slightly earlier, in others with a slight delay. The primordial gas, made up of particles at very high temperature, would bear the hallmarks of these small differences. In the regions where the Big Bang occurred earlier, the gas would have had a little extra time to expand and dilute before the Big Bang occurred elsewhere. As a result, the energy density in those regions would have been a bit lower than average. The opposite would have happened for the regions where the Big Bang was triggered later, which would have been a little denser than average. Ultimately, the structure of the original quantum fluctuations in the vacuum energy translates, after the Big Bang, into a similar structure in the distribution of matter in the universe. So, before the Big Bang, there would have been a complex geography of regions with different vacuum energy densities, and after the Big Bang, this would have been transformed into a pattern of slight variations in the density of the primordial gas.

It would be natural to think that these tiny density variations would soon disappear, just as the soup in a pot becomes homogeneous after a good stir. But another miracle of inflation now comes into play.

Inflation expands the regions occupied by quantum fluctuations so much that they are pushed far beyond the cosmic horizon. From then on, no physical process could do anything to restore equilibrium. In other words, the density variations remained engraved in the universe, frozen in time and buried in the silence of space beyond the cosmic horizon, just as happened on that fateful day in Pompeii, in 79 CE, when life suddenly stopped, trapped in the lava of Vesuvius.

After the Big Bang, the cosmic horizon overtook the expansion of space, steadily encompassing more and more regions, as shown in Fig. 10.2. The quantum fluctuations, recorded in the matter distribution and previously hidden beyond the horizon, now progressively came back into view, just as the excavations in Pompeii have gradually brought the buried city back to light following its temporary disappearance.

After the Big Bang, the matter distribution in the observable universe was thus generally uniform, albeit with small density variations. About sixty thousand years after the Big Bang, the expansion of space had slowed enough to allow the denser regions of matter to attract other matter from the surrounding space. And so began the process of gravitational collapse. Initially, this concerned dark matter accumulating around the slightly denser primordial seeds. At this stage, pressure prevented the gravitational collapse of atomic matter, which thus remained dispersed in space. Then, about half a billion years after the Big Bang, pressure was no longer sufficient to withstand the effect of gravity, and atomic matter began to precipitate within the accumulations of dark matter. The various stages in the processes leading to the formation of cosmic structure are shown schematically in Fig. 11.2.

Since the smaller regions were the first to cross the horizon, galaxies formed progressively, with the smallest first, then gradually larger ones, in a cosmic crescendo, up to clusters (groups of hundreds or thousands of galaxies held together by gravity) and gigantic superclusters (groups of clusters and isolated galaxies). At this stage, the first stars began to form within the galaxies, and the sky was at long last lit up with visible light, thus ending an eternity of total darkness. About nine billion years after the Big Bang, dark energy began to dominate, preventing the creation of new structures. This is why there are no cosmic structures larger than galactic superclusters.

The Earth is part of the Milky Way, a galaxy located in a region of space where the Big Bang occurred a bit later than average. This produced a slight

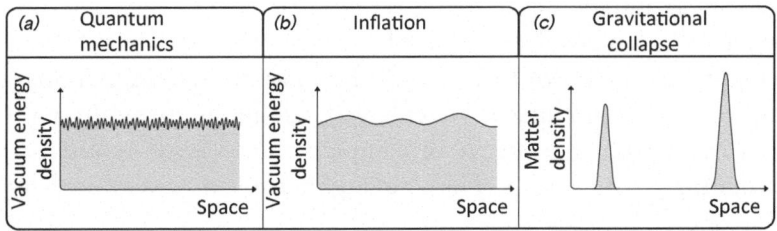

Fig. 11.2 Schematic view of the three processes that determine cosmic structure. **a** Before the Big Bang, quantum mechanics imposed random fluctuations on the vacuum energy, each roughly the size of the cosmic horizon. **b** Inflation exponentially dilated the spatial extents of these fluctuations, dragging them outside the horizon, where they remained temporarily frozen, because no physical process could affect them. The amplitudes of the fluctuations were not modified at this stage. **c** After the Big Bang, fluctuations in the vacuum energy turned into fluctuations in the matter density. The cosmic horizon began to grow, bringing the fluctuations back within the observable universe. At this point, gravitational collapse was able to amplify the amplitudes of the fluctuations, as matter fell toward regions with an initial density excess, thereby creating galactic structures

excess of matter in our vicinity, and this excess has grown by attracting matter from less dense regions nearby.

Italian railways are much more obliging than Swiss ones. I know this well, because I have the bad habit of often arriving late at railway stations, and I am always grateful when I find that the train is still there, waiting patiently for me at the platform, because it has not left on time. In the same way, humanity should be grateful for the quantum fluctuation that delayed the Big Bang in our neighborhood of the universe. Our very existence can be attributed to this random quantum jolt.

Quantum Fluctuations and Fractals

During inflation, the vacuum substance is in constant fluctuation, as quantum mechanics acts incessantly in the cosmos. The older the fluctuation, the more time there is for it to expand, driven by space dilation. After a sufficiently long period of inflation, this repeated process of quantum creation and subsequent space expansion will generate a set of vacuum energy fluctuations whose extents are equally distributed among all possible sizes. Meanwhile, the amplitudes remain almost all the same, because they are caused by similar quantum jolts. Consequently, inflation predicts that fluctuations will occupy regions of every size in a perfectly democratic manner, with no preference of large over small or vice versa. In scientific language, the inflationary fluctuations are said to have a *scale-invariant distribution*.

A scale-invariant geometric figure is one that remains the same no matter how much it is enlarged or reduced. Such figures are called *fractals*, an example of which is shown in Fig. 11.3. It is easy to find animations of fractals on the web. As the images are enlarged, the figure magically reproduces itself in a never-ending sequence of complex multicolored geometric shapes and fascinating artistic effects. The scrolling images can have quite a hypnotic effect.

Fractals are not just a mathematical oddity or a visual arts gimmick. They abound in nature, popping up among the most unexpected phenomena: the

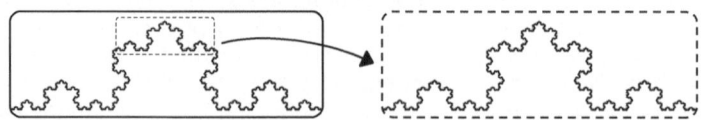

Fig. 11.3 Example of a fractal, known as the Koch curve. Enlarging a portion of the figure, we recover the original figure. The process can be repeated an infinite number of times and the figure will continue to reproduce itself

reproductive patterns of cells, the distribution of earthquakes, the branching of rivers, the structure of lightning, the rings of Saturn, the shape of Romanesco broccoli, and a multitude of other examples. Inflation teaches us that nature has also used fractals in the distribution of the quantum fluctuations that underlie cosmic structure. The universe before the Big Bang was a gigantic fractal.

If we could observe the universe before the Big Bang, focusing on a region much larger than the cosmic horizon, we would see something very similar to the simulations of fractals that can be downloaded from the web. We would see a space filled with vacuum energy, whose density varied slightly from place to place, with a structure that seemed to change over time, but almost obsessively repeating itself. The largest structures would escape from view, stretched by the expansion of space, while new structures would come into being at microscopic scales. While the universe proceeded in its unstoppable dilation, the image we focus upon would just go on repeating relentlessly in an infinite series of self-reproduction.

Observing such images of the universe before the Big Bang, there would be no way to recognize the passage of time, because everything would just go on repeating itself. There would be no past and no future, just a present filled with quantum fluctuations disappearing beyond the horizon and then reappearing within it.

But this almost distressingly stationary image of the universe before the Big Bang is only an approximation. It would prevail only if the average value of the vacuum energy remained constant over time. Instead, as shown in Fig. 9. 2, the vacuum substance evolved during inflation, albeit very slowly, and this variation would have provided something like the ticking of an imaginary clock that could be used to define a flow of time during inflation.

The correction introduced by the slow but sure variation of the vacuum substance means that the inflationary fluctuations were not in fact perfectly scale-invariant. The oldest fluctuations were generated when the vacuum energy was slightly higher than during the most recent fluctuations. Since the oldest fluctuations had more time to expand and therefore occupy larger regions of space than the later ones, inflation predicts that the largest cosmic structures should be slightly more common than the smaller ones. In short, inflation was not as democratic as scale invariance would require: large structures were slightly favored over small ones. In short, the universe before the Big Bang was almost a fractal, but not quite.

In cosmology, the distribution of structures in the universe is measured by a quantity called the *tilt parameter*. The tilt is equal to 1 if there is scale invariance, greater than 1 if small structures are favored, and less than 1 if large structures are favored. Inflation therefore predicts that the tilt parameter should be close to 1, but a little less. The most recent astronomical observations have shown the tilt to lie between 0.96 and 0.97, taking into account experimental uncertainty.

It is extraordinary that this fascinating story of quantum fluctuations, fractals, and an evolving vacuum substance should be capable of making predictions about measurable quantities in today's universe, and that the match with these measurements is so very good.

The Fossils of Radiation

Immediately after the Big Bang, matter and radiation were intimately connected. Indeed, it makes little sense to distinguish them because, at very high temperatures, matter behaves like radiation and radiation-like matter. The fluctuations inherited from the quantum phenomena that occurred during the inflationary epoch were therefore imprinted not only on matter, but also on radiation, in the form of small temperature variations.

These are the temperature variations we now observe in the cosmic background radiation, measured with very great accuracy by the Planck mission and shown in Fig. 7.2. According to the theory of inflation, the image shows the imprint left by the quantum fluctuations after being magnified by the expansion of space. This must count as the most extraordinary fossil ever discovered. Its immediate visual impact vividly conveys the impression of a cosmic fossil. However, physicists prefer to translate the content of Fig. 7. 2 into more mathematical terms, expressing it as a correlation between the temperature variations in the cosmic background and the angle over which the temperature difference is measured on the celestial vault. The result of this translation is shown in Fig. 11.4. The points are the measurements made by the Planck mission and the curve is the theoretical calculation based on the conditions in the universe at the instant of the Big Bang, as predicted by inflation.

There is no need to understand in detail the meaning of Fig. 11.4 to appreciate the astonishing agreement between theory and experiment. This result is one of the most impressive successes of the theory of inflation. It is a sensational piece of evidence in favor of the hypothesis that the universe spent its infancy in a period of rampant inflation, where space was filled only with vacuum energy and expanding exponentially.

At the time when the background radiation formed, 380,000 years after the Big Bang, the cosmic horizon corresponded to what we now see in the

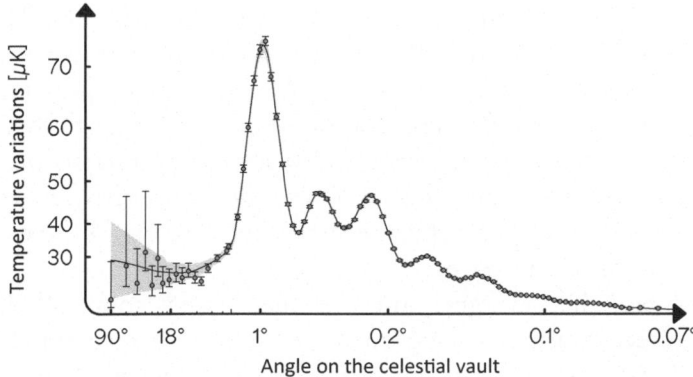

Fig. 11.4 Temperature variations in the cosmic background radiation (in millionths of a degree kelvin) as a function of the observation angle on the celestial vault, according to the measurements of the Planck mission (points with bars indicating experimental error) and according to the prediction by the theory of inflation (continuous line). The observation angle increases from right to left

sky as a region subtending an angle of about one degree. To put this in context, this is roughly equivalent to twice the angle subtended by the disk of the Moon. Points in the sky that are today more widely separated than one degree had no way of exchanging any information between the Big Bang and the moment when the background radiation was formed. This means that they reveal the original image of the quantum fluctuations produced during inflation. The approximate scale invariance of their extent explains why the temperature variations in Fig. 11.4 look roughly constant over angles greater than a few degrees, apart from a small tilt effect that favors larger angles.

At angles greater than one degree, the measurements of the temperature variations in the cosmic background shown in Fig. 11.4 are not very accurate, due to contamination from other sources. Fortunately, more precise measurements, obtained by correlating the polarization of the radiation with the temperature variations, confirm beyond any reasonable doubt the existence of fluctuations at angles greater than one degree. This result is overwhelming evidence in favor of the inflation theory because it demonstrates the existence of acausal phenomena, that is, phenomena that occur beyond the cosmic horizon. The ability to relate very distant regions of space to each other, apparently beyond the limits imposed by relativity, is indeed a distinctive feature of inflation.

For angles less than one degree, we come within the cosmic horizon, where matter and radiation have had the opportunity to interact. This gave rise to the phenomenon already encountered in Chap. 7, where the opposing effects

of gravity and pressure-induced sound waves in the radiation. The oscillations visible in Fig. 11.4 reveal the acoustic waves, or harmonics, produced
by the resonance chamber of the universe. Metaphorically speaking, they are
the cries of the quantum effects that occurred in an era prior to the Big Bang,
amplified by the expansion of space and distorted by interactions between
matter and radiation as they echoed across the primordial universe. Once
decoded, the voices recorded by the oscillations in Fig. 11.4 recount the
ancient history of the universe.

The first peak in the figure, the one that towers above all the others,
provides a direct measure of the geometry of space. By the method discussed
in Chap. 7, we can use it to measure the deformation of the imaginary
triangle drawn between our point of observation and the cosmic background.
These measurements indicate that space is flat to an extraordinarily high
degree, brilliantly confirming the prediction of inflation theory.

The comparison between the heights of the first and second peaks provides
a measure of the relative fraction of atomic matter and dark matter in the
universe, because the structure of the peaks is sensitive to the gravity and
pressure effects caused by matter. The result of the measurement is in perfect
agreement with the amount of atomic matter needed to produce the light
chemical elements in the process of nucleosynthesis, as discussed in Chap. 5.
This agreement between the data of the cosmic background radiation and
nucleosynthesis provides an excellent proof of the logical consistency of our
current understanding of the history of the universe. Nothing guaranteed
such a consistency, because primordial nucleosynthesis and the production
of the cosmic background radiation occurred at very different cosmological
times and involved quite independent physical processes.

Peak after peak, the pattern revealed in Fig. 11.4 provides new information
on the various cosmological parameters, making our current account of the
early evolution of the universe increasingly coherent and convincing. The
theory of inflation has provided a rationale for the mechanisms that generated
the Big Bang, but it has also brought together all the evidence from a whole
range of observations into a single coherent picture. The idea of inflation
offers us a logical framework able to give quantitative answers to many of the
fundamental questions about the universe that humanity has pondered since
civilization began.

The Fossils of Geometry

As we have seen, the universe is not just matter and radiation. Geometry is also a fundamental feature. According to Einstein's relativity, the geometry of space–time is not a passive structure, but a dynamic entity that changes and reacts to what is going on in the universe. It therefore makes sense that geometry should be affected by quantum fluctuations, just like matter and radiation.

Inflation has made the geometry of space almost perfectly flat, but quantum mechanics has introduced its inevitable deformations. Just as matter fluctuations give rise to galactic structures and radiation fluctuations correspond to temperature variations in the cosmic background, so the fluctuations in the geometry of space produce a potentially measurable physical effect. Since we perceive the geometry of space as the force of gravity, its fluctuations should manifest themselves as gravitational waves that propagate through the universe.

More precisely, the theory of inflation tells us that, at present, the universe must be populated by gravitational waves characterized by constant amplitude, proportional to the original energy of the vacuum, and an almost perfectly scale-invariant distribution of frequencies. The passage of gravitational waves slightly deforms space. Their effect, although too weak to be detected with current instruments, could be measured in the future by more sensitive experiments.

There is also an indirect way to detect these fluctuations in the geometry. The collision between electrons and radiation in the primordial plasma would have left the latter polarized. When a gravitational wave passes through, the polarization of the radiation takes on a characteristic form which remains imprinted on the cosmic background radiation like a signature on a sheet of paper. This signature of the gravitational wave is a fossil trace of the geometry.

Experiments currently measuring the cosmic background radiation are hunting for the signatures left on the polarization structure by gravitational waves, but no trace has yet been found. In 2014, the BICEP2 experiment, located in Antarctica, announced a discovery, but they later had to retract when other scientists understood that the data showed only a contamination due to galactic dust, capable of distorting the image of the cosmic background radiation and mimicking the effect of a gravitational wave.

The search for primordial gravitational waves is in full swing. It will be no easy matter to make such a detection, because the signal could just be too weak. However, if these waves were identified, it would be of outstanding importance. For one thing, this measurement would provide a further piece of evidence in favor of inflation, and for another, it would tell us the energy of the inflationary vacuum. We would thus obtain a key piece of information about the microscopic characteristics of the vacuum substance that gave birth to the Big Bang.

12

Parallel Universes

The only difference between me and a madman is that I'm not mad.
Salvador Dalí

In *De rerum natura*, Titus Lucretius Caro, following the teachings of Epicurus, imagined the universe as a huge ensemble of atoms scattered throughout an infinite empty space. The atoms cluster together and dissipate through random motions (*clinamen*), creating and destroying all things, without the need for divine intervention. In his view, the immortality of the soul was a nefarious belief (*tantum religio potuit suadere malorum*, religion can induce such heights of evil) and, after death, the atoms of our body scatter and recombine in other forms.

Based on his atomistic vision of the universe, Lucretius deduced that there must exist many other worlds similar to our own, yet different: "And now, if the number of atoms is so vast that the whole age of human beings would not be enough to count them, and if the same force and nature remains capable of aggregating atoms everywhere in the same way that they have aggregated here, we must admit that there exist, in other regions of space, other lands and different races of men and species of animals."

In *On the Infinite Universe and Worlds*, Giordano Bruno argued that the universe is infinite and that all the stars in the sky are similar to the Sun, with planets orbiting around them, similar to the seven planets known at the time: "There are therefore innumerable suns, and there are infinitely many earths, which similarly revolve around those suns; as we see these seven orbit this sun near us. [...] We see the suns, which are the largest, indeed very large bodies, but we do not see the earths, which, being much smaller bodies, are invisible."

In a letter to his brother, Cicero expressed his appreciation of *De rerum natura*, while the idea of infinitely many worlds brought much less success for Giordano Bruno, who ended up being burnt at the stake in Campo de' Fiori in Rome. The eccentric and mystical ideas of these thinkers went against the trend, but the curiosity for what might lie beyond our own world is irresistible.

As far as our astronomical instruments have been able to reach, namely, within the cosmic horizon, the universe has revealed its extraordinary uniformity. Inflation suggests that the universe should continue to be uniform well beyond the current cosmic horizon. But how far does the universe remain uniform?

It may seem that such a question goes beyond the boundaries of scientific exploration. How could we investigate reality beyond the cosmic horizon, or beyond everything that is humanly measurable? The answer must be sought in the deductive method, which combines experimental measurement with logical reasoning, allowing us to penetrate territories inaccessible to direct observation and telling us something about what the universe might be like beyond—indeed, much further beyond—our cosmic horizon. What we shall find is a truly surprising story. It is a great pity that Lucretius and Giordano Bruno cannot be with us today to hear it.

Eternal Inflation

When we discussed inflation, we began by considering a region of space that was within the cosmic horizon at a time well before the Big Bang. If we now want to understand the large-scale structure of the universe, we need to broaden our view. We must consider a portion of space that contains everything that exists, hence extending much further than the horizon, perhaps even infinitely far. We must fly above what is humanly observable and imagine ourselves as supernatural creatures capable of seeing the universe in its boundless entirety.

The space of this immense universe can be divided into a huge number of patches as large as the horizon. These will be causally disconnected from one another, because the limit on the speed of light prevents them from exchanging any form of information or influence. Each of these regions will have its own independent history, just as the peoples of the Americas and

Europe went about their affairs without influencing each other until the arrival of Christopher Columbus (or perhaps the Vikings).

In each patch, the vacuum substance follows its own evolution, gradually descending the slope of the potential, as explained in Chap. 9. Since the initial conditions for the value of the vacuum substance and its speed are generally different from one patch to another, the vacuum energy will vary randomly between different regions of the universe.

Now one might think that, since the vacuum substance evolves toward the condition in which the Big Bang is triggered, inflation will sooner or later have to end everywhere in the universe. In other words, if one is patient enough, one will see the entire universe undergo the Big Bang and transform into a hot mixture of particles. But this reasoning is mistaken because it does not take into account two factors that completely change the situation: quantum mechanics and the expansion of space.

Because of quantum mechanics, the cosmic evolution of the vacuum substance does not follow a uniform trajectory along the profile of the potential. Referring back to the analogy in Fig. 9.2, it is as if the ball moving inside the bowl were continuously subject to slight, random vibrations. The ball does not roll uniformly inside the bowl, but moves in fits and starts, sometimes jumping up, sometimes jumping down. These random jolts are the effect of quantum fluctuations.

In the end, the vacuum substance moves rather like a reeling drunkard, staggering down a slope, sometimes taking a couple of quick steps downhill, sometimes taking one back up. All in all, the vacuum substance does proceed toward the condition where the Big Bang is triggered, but its path is subject to chaotic fluctuations, which sometimes lead it away from its inevitable destination.

In regions of space where the Big Bang has occurred, inflation stops. In the meantime, regions where the Big Bang has not yet occurred continue to expand at the rate predicted by the theory of inflation. Because of this expansion, two opposing effects are at work in the universe: on the one hand, some regions stop inflating after the Big Bang, and on the other, the regions that go on inflating grow much faster than the others.

An example can help us to understand what happens to the universe when it is subject to these two opposing tendencies. Consider a region of space filled with the vacuum substance at some instant of time. Suppose that, for each lapse of time equal to the inflation doubling time, the vacuum substance reaches the condition for the Big Bang in a quarter of any given space region. In that quarter, inflation thus comes to an end. In the remaining three quarters, the vacuum substance is still wandering around on the potential profile, without having reached the bottom. After a further doubling time, the inflating regions double their size, while inflation ends with a Big Bang in a quarter of each of those regions. And so on after each subsequent doubling time.

The situation is illustrated in Fig. 12.1, in the simplified case where space has only one dimension. The figure shows snapshots of the initial region taken at successive time intervals equal to the inflation doubling time. The gray boxes show the regions of the universe still inflating, and the white ones the regions where the Big Bang has occurred, where space has therefore ceased to expand exponentially.

The figure illustrates the two opposing effects that govern the history of the universe. The first is the evolution of the vacuum substance which, by reducing the vacuum energy, ends up triggering a Big Bang (hence converting gray boxes into white ones). The second is the expansion of space, which greatly amplifies the inflationary regions (i.e., the gray boxes), but not those where the Big Bang has occurred (the white ones).

The fate of the universe is decided by which effect dominates. If the average time required by the vacuum substance to trigger a Big Bang is short compared to the doubling time, Big Bangs will convert all available space and inflation will die out. But if the opposite is true (as in the example in Fig. 12.1), the volume of inflationary space will increase over time, rather than decrease. In this case, somewhere in the universe there will always be some region where inflation is still tirelessly boosting the volume of space, and *inflation will be eternal.*

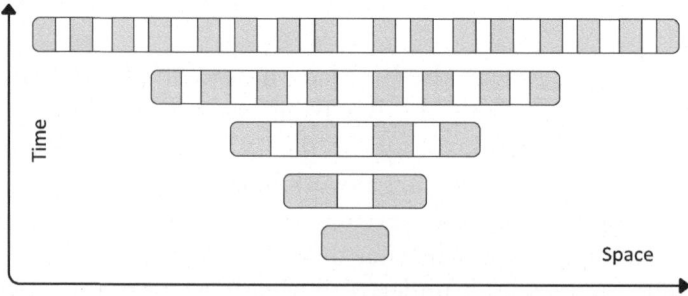

Fig. 12.1 Evolution of a region initially in the inflationary phase, viewed at successive time intervals equal to the inflation doubling time. The space here is assumed to be one-dimensional. After each time interval, each inflationary region (indicated in gray) doubles, while a quarter of it undergoes the Big Bang and transforms into a region (indicated in white) occupied by hot gas. Due to the expansion, the inflationary regions (in gray) never completely disappear, whence inflation persists eternally in the universe

The situation is analogous to what happens in an animal population. The mortality rate corresponds to the probability that a Big Bang occurs, that is, that regions of space end their inflationary life. The birth rate corresponds to the rate of expansion of space, that is, the rate of growth of inflationary regions. The survival or extinction of the animal population is decided by whichever wins out between the mortality rate and the birth rate.

The particle nature of the vacuum substance is still unknown so we cannot establish with certainty whether or not inflation is eternal. Moreover, the eternal stability of de Sitter space remains controversial. Despite many unresolved issues, the hypothesis of eternal inflation is fascinating enough to deserve further exploration.

The mix of inflation and quantum mechanics gives us an insight into the possible structure of the universe on very large distance scales, well beyond the cosmic horizon. This structure may well be much more complex than we might have imagined. Our exploration leads to the discovery of a new world: the *multiverse*.

The Multiverse

An astronaut can rise above the two-dimensional space of the Earth's surface and get an overview that would be impossible for a human being who lives at a certain point on the surface and whose view is limited by the horizon. From a spaceship in orbit, the astronaut can observe the whole of the Earth's surface and admire the larger structures of continents, islands, and oceans.

Now imagine doing the same thing for the universe. Imagine being a supernatural creature, call it Big Brother, able to rise above three-dimensional space and observe the entire universe with a view wide enough to cover distances far greater than the cosmic horizon. What image would you have before your eyes?

Eternal inflation tells us that the universe would appear as an immense vastness of space in rapid expansion. This expansion would not be uniform everywhere. Some patches of space would be expanding faster than others. Those that expand faster are born less frequently than others, because they require more vigorous quantum fluctuations, capable of raising the vacuum substance further up the potential. On the other hand, because they expand faster, those patches will cover larger regions, thus making up for their initial rarity. The patches will change continuously, offering Big Brother a thoroughly dynamic view of the universe, like a stormy ocean crisscrossed by waves of all sizes. But the story doesn't end there.

Here and there in this vast ocean of space, bubbles suddenly form, where the vacuum energy is instantly converted into thermal energy, creating a gas of particles. These are Big Bang events. Compared to the full vastness of space in impetuous expansion, they involve seemingly insignificant islands of space. New bubbles are born all the time in different parts of the universe, each following a new Big Bang. According to this picture, the cosmos is like an ocean in rapid and heterogeneous expansion, dotted with islands that emerge from empty space with every new Big Bang event.

Each of those islands is a new universe that will follow its own evolutionary course, completely oblivious of the storm raging outside its borders. Although they are created incessantly, the islands will never fill the entire ocean, because the space between them is expanding wildly, making room for new empty space and new islands.

Our supernatural Big Brother is witness to an impressive spectacle of space in tumultuous expansion, where new universes come bursting into existence from recurring Big Bangs, like some phantasmagoric firework display. He is observing the *multiverse*.

From its etymology (*universus* = *unus* + *versus*, one + facing in the same direction, that is, the whole), the word 'universe' is usually taken to mean everything that exists. But this word could never suffice to describe the grandiose reality that lies beyond the cosmic horizon of the visible world, so physicists have invented the word 'multiverse' to convey the idea of an immense multitude of isolated universes, immersed in an empty space filled with the vacuum substance. In physics, the multiverse is the ensemble of all the individual island universes and the empty space that separates them.

The multiverse presents us with a completely different picture of the cosmos from the one built on the assumption that nearby astronomical observations provide a faithful portrait of the entire physical reality. The multiverse shows us that something different and surprising lies hidden beyond the cosmic horizon. The idea that the universe on a large scale is flat and uniform is just an illusion derived from our limited view. Not only do we live inside a bubble of space, but we cannot even see beyond the cosmic horizon, which is itself much smaller than the bubble. We are like castaways in the middle of the sea, prisoners of the illusion that the flat surface of the water is all there is, unaware of the lands and mountains that rise beyond the horizon.

At much larger distance scales than our bubble, the cosmos reveals itself to be anything but a simple uniform expanse. Instead, what we have is a complex structure composed of a heterogeneous expanding void, dotted with universes crammed with matter. The picture that emerges is curiously similar to the image of a sky dotted with galaxies. And this is no accident, because mathematical calculations reveal that the size distribution of the island universes is once again fractal. The same abstract form we encountered in the distribution of quantum fluctuations which gave rise to the galactic structures is repeated in the distribution of island universes within the multiverse. It is almost as if nature loves to replicate itself at different scales, producing every conceivable variation in an endless game of mirrors.

Eternal inflation undermines the significance of the Big Bang, which is no longer a special event that happened only once in cosmic history. On the contrary, Big Bang events occur all the time, like raindrops falling on the ground during a storm. At this very moment, a Big Bang is under way somewhere in the multiverse. And Big Bangs will continue to happen forever more.

In its eternal form, inflation is not confined to the remote past of our universe. It is an enduring feature of physical reality. At this very moment, regions of space far away from us are undergoing dizzying inflation. Inflation will always be going on somewhere in the multiverse, and this for all eternity.

The Ultimate Copernican Revolution

If I had lived in ancient times, I would have defended the thesis that the Earth sits in the very center of the universe. After all, everything seems to point in that direction. The Sun's chariot crosses the sky every day, following an arc from east to west. The stars move as the seasons go by, but the distances between them remain constant, so it looks as though they are glued onto a background that rotates around us. The constellations fixed in the firmament pick out symbolic figures, carriers of mysterious messages. The stars look like immutable features of some inexorable sidereal mechanism, while life on Earth is in a continuous state of change. The human soul is immeasurably more articulate than these mute celestial bodies. All the evidence would have encouraged me to believe that humanity occupies the center stage of the universe. It took time and a growing trust in the scientific method to understand that things were not like this at all.

The Copernican Revolution was a major intellectual upheaval. It did not just reverse the order of the celestial orbits. It upset the very foundations of human thinking about the principles that regulate the universe. By demoting the Earth from the cosmic center and making it revolve around the Sun, Copernicus started a process that has proved unstoppable.

The Copernican Revolution continued when we realized that the Sun is only one among myriad other stars and that the Solar System rotates around the center of the Milky Way galaxy, completing its orbit in about 230 million years. It continued when Hubble's observations confirmed that the Milky Way is just an island of stars, similar to the other thousand billion galaxies now known to populate the observable universe. It was subsequently discovered that, in the universe, the atomic matter that makes us up is five times less abundant than dark matter, which in turn has an energy content almost three times lower than what is now called dark energy. In short, it would be difficult today to argue that humanity occupies a privileged position in the universe.

The multiverse pushes the Copernican Revolution to the extreme. Not only are the Earth, the Solar System, and the Milky Way anonymous individuals among a huge multitude of others, but, according to the idea of eternal inflation, the same can even be said of our whole universe. Our observable universe is just a grain of sand, similar to countless others, abandoned to the tumult of a vast ocean.

Perhaps there is even more truth in Giordano Bruno's idea than he could ever have imagined. His infinite worlds might not simply be other stars with other planets orbiting them, but other universes, distinct from ours, born at different times in the past and yet to be born in an eternal future.

After being forced to kneel to hear his sentence, Giordano Bruno addressed his accusers with the words: "Perchance you who pronounce my sentence are in greater fear than I who receive it" (*Maiori forsan cum timore sententiam in me fertis quam ego accipiam*). He knew that no court could ever curb the human desire to push back the frontiers of knowledge.

Is the Universe Infinite?

Philosophers throughout the ages have been divided on the question of the size of the universe: some believed space to be infinite, others that it was finite, and others still it is hard to know what they were saying. Yoko Ono chose a middle way when she composed her album *Approximately Infinite Universe*. Eternal inflation offers a surprising answer to this age-old question about the infinitude of the universe. It turns out that the answer may be more in harmony with the music of Yoko Ono than with the opinions of the philosophers.

To address the question, consider a single island universe generated in the multiverse. Figure 12.2a shows the island in a space–time diagram where, as usual, space is represented by a single dimension. From the point of view of Big Brother—the supernatural being with an overview of the whole multiverse—each island is born in a small region of space at a given instant of time, identified as the Big Bang for that particular island universe. As time goes by (proceeding from bottom to top in the figure), the island grows larger, due to the expansion of space. Since the island universe is born in a limited region, its space is necessarily finite at the moment of the Big Bang. The cosmic expansion amplifies this region, but it can never transform it into an infinite space during any finite time. Hence Big Brother asserts with absolute conviction that, according to his version of the facts, *the space of each island universe is finite.*

Big Brother's categorical assertion refers to the space of each single island universe and not necessarily to the space of the entire multiverse, which may well be either finite or infinite. In other words, we do not know whether Fig. 12.2a extends infinitely far to the right and left. However, we do not need to know this to continue our reasoning and reach valid conclusions.

Relativity teaches us that reality can look very different to different observers. Our particular Big Brother has the privilege of observing the multiverse as a whole, viewing from outside of space. However, we can imagine several different Big Brothers in relative motion to each other. Each will see the island universes born in a different chronological order. Each will thus

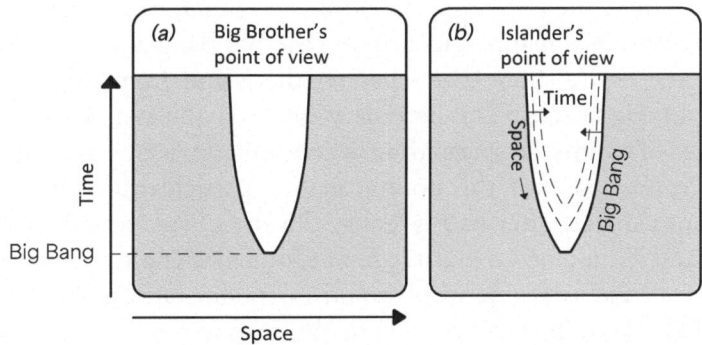

Fig. 12.2 An island universe (white region) born within the ocean of inflationary space (gray region), in a space–time diagram with a single spatial dimension. **a** Point of view of Big Brother. The island universe is born at the instant identified as the Big Bang and then expands. **b** Point of view of an inhabitant of an island universe. The Big Bang is identified as the event that occurs throughout the region of space corresponding to the shore of the island, and time is measured as the distance from the shore. Dashed lines indicate regions of space at the same instant of time, measured since the Big Bang

claim to have seen a different reality and none of them will be lying. The fact that they can tell different stories is a consequence of relativity.

Among all the possible perspectives that exist, one is particularly interesting: the point of view of an inhabitant of an island universe. Naturally, this is a limited perspective, because an islander cannot observe the entire multiverse, but it is nevertheless relevant, precisely because we humans are inhabitants of an island universe.

There is no absolute way to define time in the complex heterogeneity of the multiverse as a whole, but within a single island, there is indeed a clock on which every inhabitant can read the time. This is the clock of cosmic evolution.

The general uniformity of matter and radiation within the given island at each instant of time, which are guaranteed by inflation, allows us to attribute a universal meaning to the passage of time, that is, one that does not depend on which inhabitant reads the clock. Relative to this scale of time, everywhere on the island, the first composite atomic nuclei begin to form about twenty seconds after the Big Bang, while the gravitational collapse of dark matter begins some 60,000 years later, and the cosmic background radiation is emitted 380,000 years after the Big Bang; and so on until all the nuclear fuel in the stars is exhausted. This is the timeline laid out by the evolution of the universe.

Even those islanders living at distances well beyond the cosmic horizon, who cannot therefore communicate with each other, can all agree to define time in this universal manner, identifying the Big Bang as time zero. It corresponds to the entire shore that separates the island from the ocean in the schematic of Fig. 12.2b. The boundary between the island and the ocean is the locus of points in space–time where inflation ends and the vacuum energy is converted into the boiling gas of particles that constitutes the universe immediately after its Big Bang. On the island, time flows from the Big Bang and simultaneous events are not located along the same horizontal line of Fig. 12.2a, as Big Brother would maintain, but on the dashed lines in Fig. 12.2b. Time flows in sync with the cosmic evolution and, when we venture further into the island, we explore an ever older universe.

For Big Brother, the Big Bang occurs in a particular place. For the islander, it occurs at a particular moment, but simultaneously throughout space.

The shore of the island in Fig. 12.2 is not a closed curve. This is because the theory of inflation predicts that the island universe has a flat geometry, so the effects of expansion will never completely come to an end. Moreover, if inflation is eternal, the multiverse will exist forever and time will have no end. Since the shore represents the whole space of the island universe at the instant of the Big Bang, we arrive at the extraordinary conclusion that the universe was born in an infinite space. This is the inevitable conclusion from the facts as viewed by the islanders, who have the same perspective as we humans: *the space of each island universe is infinite.*

This conclusion may seem paradoxical. Inflation has created an infinite space within a speck of the multiverse enormously smaller than a proton. Indeed, it can create an infinite number of island universes of infinite size within a multiverse with a finite space. But doesn't this violate the laws of physics? And how is it possible for a Big Bang to create an infinite region of space in a finite lapse of time?

These questions do not lead to paradox. Instead, they reveal the inherent magic of eternal inflation. And as with anything that seems like magic, there has to be some kind of trick. In this case, the trick is relativity, which replaces absolute space and absolute time by a mixture of the two: space–time. Reality in space–time is of course absolute, but it looks different to those who measure space and time in different ways.

Imagine cutting space–time into slices like a ham. Each slice corresponds to space at a given time, while the ham itself represents the whole of space–time. The way space–time is sliced up depends on how time is measured. Observers in different places or with different motions each have their own perception of time which corresponds to a different way of slicing up space–time. The spatial slices therefore look different to different observers: for some of them, space is infinite, and for others, it is finite. There is nothing paradoxical about this. It is just one of the many strange things we have to deal with in a relativistic world.

At each instant of time, Big Brother sees the space of the island universe as a horizontal section of Fig. 12.2a, so the universe is finite. But the islander, at each instant of time, sees space as a section along a dashed line of Fig. 12.2b, so the universe is infinite. The trick lies in the way space and time are mixed together. In eternal inflation, as the name suggests, time is infinite in the future. The islander interprets the infinity of time as an infinity of space.

Yoko Ono was right if, by 'approximately,' she meant that the infinity of our universe is just a mirage as it would appear to a supernatural being that could observe the multiverse as a whole, hovering outside of space.

The Nightmare of Infinite Replicas

There is a tower block in Geneva where the inner walls of the lift are completely covered in mirrors. When I go up to the higher floors, my gaze inevitably turns toward the wall, where I see an endless row of images of myself, reflected back and forth by the two opposite mirrors. These infinite replicas make me feel rather uneasy, but turning the other way does no good. The opposite wall shows another endless sequence of these images, staring back at me uncomfortably. So, even when I have to go up to the top floor, I prefer to take the stairs.

This silly discomfort of mine is nothing compared to what we might feel when we think about eternal inflation. Let us focus on the single island universe in which we live and divide it into patches as large as the current cosmic horizon, that is, with a diameter of 93 billion light-years. Since we now know that the space of our island universe is infinite, an infinite number of patches will be needed to cover it. These patches are just like parallel universes, in the sense that they exist simultaneously but have no way of communicating with each other. Since they were born from the same Big Bang and thus have a common origin, these parallel universes are almost identical.

So far, the story seems rather dull: an infinite sequence of parallel universes, photocopies of each other, like the images reflected in the walls of the lift. But quantum mechanics pulls off an unexpected *coup de théâtre*.

If it weren't for quantum mechanics, all the patches would be perfectly homogeneous and identical. Instead, quantum fluctuations add small variations to the energy of the vacuum that fills the space before the Big Bang. As discussed in Chap. 11, these variations are the seeds that lead to gravitational collapse and end up forming the galactic structures present in the sky today. Since quantum fluctuations are random, each parallel universe follows its own evolutionary history, and when the fluctuations are amplified by gravitational collapse, these histories can progress in very different ways. Parallel universes have an identical global structure, determined by their common inflationary origin, but they can differ significantly in terms of galaxies, stars, planets, and forms of life.

The second ingredient introduced by quantum mechanics is the blurring of physical quantities due to the intrinsic uncertainty postulated by the theory. This indeterminacy implies that the number of possible evolutionary histories of a cosmic patch is not infinite, because histories that differ only infinitesimally are necessarily indistinguishable, according to quantum mechanics.

The number of possible evolutionary histories of the quantum system can be calculated. The result is a huge number, but finite.

The third ingredient also has to do with quantum mechanics, and in particular its intrinsically probabilistic nature. Each evolutionary history has a certain probability of occurring, so it will certainly come about somewhere in the infinite set of cosmic patches. According to quantum mechanics, everything that is not impossible will sooner or later happen. From this point of view, quantum mechanics obeys the same principle as Roman law: *ubi lex voluit dixit, ubi noluit tacuit* (where the law willed, it spoke; where it did not, it was silent). Translated into scientific language, this liberal interpretation of reality means that any physical process not expressly prohibited by a fundamental law has a certain probability of happening in the quantum world.

Put together, these three ingredients lead to a surprising but inevitable conclusion. Since there is a finite number of cosmic histories for an infinite number of parallel universes, and since each cosmic history must necessarily occur, then *every possible cosmic story must necessarily happen in some parallel universe and repeat itself in infinitely many parallel universes*. What underlies this disconcerting result is the rather mysterious concept of infinity, which can hide any number within itself infinitely many times.

This is more troubling when we consider what it means for human life. Our very existence is proof that life is a possible consequence of cosmic history. But then, according to eternal inflation, it must have occurred not just once, but an infinite number of times in parallel universes. Among these infinitely many possibilities, there will be an unlimited number of exact copies of each of those who are now reading or writing this book. Or copies of those who did not casually throw away a youthful love, or who have met the love of their life in an appointment that we ourselves missed by five minutes. In other words, Napoleon lost or won the Battle of Waterloo an infinite number of times, Lars Porsena defeated the Romans and an Etruscan dialect is now spoken in Italy, Renzo and Lucia died of the plague, while Romeo and Juliet got happily married.

In his short story *The Library of Babel*, the Argentine writer Jorge Luis Borges tells of a library that contains all possible books, composed of every possible permutation of any number of letters. All the stories that have ever been written or will ever be written are already contained there, along with all possible philosophical treatises and an endless number of sequences of letters devoid of meaning, or perhaps written in ancient idioms now forgotten, or in languages not yet invented. Every single island universe is a Library of Babel, on whose shelves are stored countless parallel universes.

"*Nihil sub sole novum,*" we read in Ecclesiastes: there is nothing new under the Sun. In the light of eternal inflation, the biblical statement takes on a much broader significance. There is nothing new or original anywhere. Every thought that comes to our minds has already been thought an infinite number of times in parallel worlds. Every idea we ever contemplate has already been examined and perhaps discarded for good reasons that still elude us. Even eternal inflation is a theory that has already been invented an infinite number of times.

Ever since I understood the full consequences of eternal inflation, entering the lift in that tower block in Geneva has ceased to bother me. There are far worse nightmares than seeing oneself replicated in a mirror. Studying eternal inflation has thus saved me the need to take the stairs.

13

The Complexity of the Multiverse

We adore chaos because we love to produce order.
M. C. Escher

In *Alice's Adventures in Wonderland* and *Through the Looking-Glass*, Alice discovers new worlds where the natural order of things is radically changed. By biting into one side of the mushroom on which the Hookah-Smoking Caterpillar sits, Alice can become tiny, while a bite on the opposite side transforms her into a giant. In the March Hare's house, time flows differently, while the Mad Hatter's clock has stopped at six o'clock, so that it's always time for tea. Playing cards and chess pieces come to life. The Red Queen runs as fast as she can, yet always remains in the same spot. The fat twins Tweedledum and Tweedledee try to convince Alice that she is only an imaginary figure in a simulated reality that lives in the dreams of the Red King, asleep at the edge of the chessboard.

The worlds that Alice visits are different from those imagined by Lucretius. The atoms in *De rerum natura* can combine into different forms, but they conform to the same natural order. The physical laws in Wonderland, on the other hand, contradict those governing our own reality. They are so unreal that Lewis Carroll ends his story by waking Alice from the dream that allowed her such wonderful adventures, and concludes the book with the words: "Life, what is it but a dream?" But what if these unreal worlds were not just a dream? Could there exist parallel universes where the physical laws are so radically different?

The infinitely many parallel worlds hidden within a single island universe that we encountered in the previous chapter are already enough to make our

G. F. Giudice, *Before the Big Bang*, Copernicus Books, https://doi.org/10.1007/978-3-031-69933-7_13

heads spin. The unlimited copies of every admissible reality are troubling enough to awaken existential nightmares. Yet I will now tell of an even more disconcerting aspect of the multiverse.

The physicist Alexander Vilenkin is one of the pioneers of the idea of eternal inflation. Reserved but affable, he is a man of exquisite courtesy. In Soviet times, he was a student in Kharkiv—the city now sadly known for the atrocities committed by the invaders of Ukraine—when he was approached by KGB agents who asked him questions about the political opinions of another university student. Vilenkin refused to cooperate and soon found that his enrollment in doctoral courses had been unexpectedly suspended. The reasons were never communicated to him.

He was thus forced to look for odd jobs and ended up as night watchman at the local zoo, studying physics on his own during the day. Thanks to a program for Jewish emigrants, he managed to escape from the Soviet Union and finish his studies in the United States, where he is now a professor at Tufts University.

When Vilenkin proposed the theory of eternal inflation in 1983, the idea did not gain much traction. At that time, inflation was still only an abstract hypothesis, and eternal inflation seemed an abstruse fantasy built on an abstract hypothesis. The only recognition Vilenkin could get was from a nostalgic fan when he pointed out that Elvis Presley must still be alive somewhere in the multiverse, among the infinite replicas of parallel worlds. But not even the ruse of bringing back the King of Rock 'n' Roll was enough to attract attention in the physics community.

Things changed when it was noticed that string theory might have revolutionary consequences for the reality of the multiverse. To understand the scope of this revolution, we must first say a few words about string theory.

The Quanta of Space–Time

General relativity offers an elegant description of gravity and all experimental tests so far have been successful. Nevertheless, the theory has the symptoms of an illness indicting that it cannot be universally valid.

Common experience suggests that matter, space, and time are continuous media that can be divided into an infinite number of parts. Quantum mechanics, on the other hand, teaches us that matter is not at all continuous and, at small distances, breaks down into *quanta*—which are commonly called particles. Matter is like a computer image: it appears continuous when viewed at normal resolution but, sufficiently magnified, it fragments into distinct pixels.

In general relativity, space and time are described as continuous media. Since space and time are, like matter, dynamic entities that react to physical phenomena, it should not be surprising that the geometry of space–time, when it enters the regime of quantum mechanics, also breaks down into quanta. This happens at distances a hundred billion billion times smaller than a proton. These distances are almost too small to imagine, but mathematics comes to our aid. On such infinitesimal length scales, space–time breaks apart into quantum entities and it no longer makes sense to describe it as a continuous medium. We still do not know what will replace general relativity when space–time enters the quantum regime, but we know for sure that Einstein's theory will be superseded by something radically different. Pending its discovery, the unknown theory has been given the provisional name of *quantum gravity*.

The difficulty in formulating quantum gravity arises from paradoxical infinities that appear in calculations as soon as we try to reconcile general relativity with quantum mechanics. Since the origin of the problem is the point-like nature of particles, one possible remedy is to assume that particles have an internal structure, and the simplest extension of a zero-dimensional point is a one-dimensional line. This idea is the starting point for *string theory*.

In string theory, the fundamental elements of reality are one-dimensional extended objects, similar to very thin strings, free to wrap themselves around each other in complicated tangles. The vibrations of these strings correspond to different states of particles, just as the strings of a violin emit sounds at different frequencies, vibrating in one of the infinitely many possible harmonics. Unlike the strings of a violin, these strings are not necessarily fixed at the ends, but can flit about in space or close up to form loops.

The broad interest in this new conception of reality stems from the discovery that gravity is an inevitable consequence of string theory and can be described in a way that is perfectly consistent with the dictates of quantum mechanics. In short, string theory could be the long-awaited answer to our quest for quantum gravity. There are other points in its favor. For example, it allows an exact count of the quantum states of black holes and contains within it everything needed to describe the forces between the known elementary particles. So, there are excellent reasons to believe that string theory plays a fundamental role in the natural order of things.

Since the last decades of the twentieth century, theoretical physicists have made a huge effort to understand the structure of string theory. There was a feeling of being close to the discovery of a new archetype in nature, from which all physical laws could be deduced. It was the dream of a final unification, a dream in which our physical reality is the sole possible consequence of a single fundamental principle. The only obstacle was to understand the complex mathematics required to formulate the theory and the reward would have been an unprecedented prize. Some of the best minds in theoretical physics immediately set to work with enthusiasm and dedication.

The mathematical structure of string theory proved to be a tough nut to crack. Despite tremendous progress, no one was able to unravel the equations and find the one possible solution describing the entire physical reality in a single stroke. Some began to suspect that these failures might not be due to a lack of mathematical inventiveness, but mask a hidden property of strings.

Any particle theory contains *physical parameters*, that is, numbers that determine its characteristics: the strength of the forces between the constituents, the masses of the particles, the energy of the vacuum, and similar. These parameters cannot be calculated in particle theory and must be derived from experimental measurements. One of the most attractive aspects of string theory is that all physical parameters are in principle calculable. There are no free parameters and all the fundamental constants of nature are determined by the theory itself. The problem is that, to find the values of these parameters, we need to know the state of the system, and this requires a full solution of the equations of string theory.

A similar situation is illustrated by water, which exists in three states: liquid, solid, and gas. It is always the same substance, but ice looks very different from water vapor. Similarly, string theory is unique, but it can manifest itself in different states, each corresponding to different values of the physical parameters.

Unlike water, which can only assume three states, string theory is a gigantic chameleon, capable of changing shape and appearance in a colossal number of ways. This number is enormously larger than the number of stars in the observable universe; larger than the number of legal positions of the pieces in chess (which are thousands of billions of billions of times more numerous than the stars, giving us an idea of how phenomenal the supercomputer Deep Blue was in defeating world champion Garry Kasparov). Enormously larger even than the number of possible positions in the game of Go (which is, in turn, vastly greater than the number for chess). In short, the number of possible forms that the chameleon of string theory can assume is so gigantic as to defy all imagination.

Completely different physical realities are just different faces of the same structure: string theory. And there is one further surprise: the different states of the multifaceted string theory are determined by purely fortuitous circumstances, impossible to predict with any accuracy.

It is at least ironic that a theory conceived to realize a single inevitable reality should end up predicting a huge range of alternatives. The theory that was supposed to reveal the final unification sanctioned by a single universal principle leads us into a labyrinth of heterogeneous worlds, all equally possible. This surprising result is characteristic of what happens in scientific research. Science proceeds according to the canons of logic but it is not uncommon to end up in quite the opposite place to the goal originally set.

The Secret Life of the Multiverse

It is easy to imagine the disappointment felt by theoretical physicists upon discovering this huge panorama of possibilities, when they believed they had the ultimate key to unlocking the secrets of nature. But here eternal inflation comes into play and once again completely changes the stakes, transforming a seemingly disheartening result into a new and fascinating opportunity.

If indeed the theory that describes the constituent elements of matter can manifest itself in unthinkably many different ways, each depending on fortuitous circumstances, then each of the island universes will turn out to be a different kind of world. The physical parameters of each island universe will be determined randomly, as if chosen by the roll of a die. It is a very special roll, because the die in this case doesn't have six faces, but a number so colossal as to make one's head spin.

Each island universe, with its physical parameters determined by the roll of the die, is full of particles with different masses and properties, governed by different fundamental forces. It possesses a different vacuum energy, and can even differ in the number of dimensions of space. In short, each island universe contained in the multiverse distinguishes itself from the others by its own very special reality, as if all possible fantastic stories ever imagined were recounted in the same book.

String theory suffuses the panorama of the multiverse with new colors. According to this theory, the island universes that populate the multiverse are not an unlimited army of almost identical clones. Each realizes a unique and highly original world. Ultimately, the island universes look much more like the surreal worlds of Lewis Carroll than those imagined by Lucretius, made up of atoms that combine in different ways, but always respecting the same rules.

The multiverse is the antithesis of Einstein's conception that fundamental principles must give rise to a single logically coherent universe. The multiverse rejects the monotheistic approach to physics, in favor of a heterogeneous and harmonious vision, where different realities can coexist.

It should not be thought that the multiverse describes the ultimate cosmic chaos, able to accommodate any kind of physical manifestation. The theory that populates the islands of the multiverse, whether it be string theory or a different form of quantum gravity, provides the framework for the fundamental laws. However, as the physical parameters vary, objective reality can radically change in appearance. This protean character of the multiverse opens the way to a new conception of the natural order.

The Elusive Boundary Between Simplicity and Complexity

The aim of physics is to identify a pattern behind natural phenomena and encode it in mathematical equations, which we call physical laws. This process is often guided by a search for simplicity, even though the meaning of simplicity is vague and subjective. We perceive simplicity according to intuitive standards. Any theoretical physicist would be ready to assert that general relativity fulfills the requirement of simplicity, even though its mathematical formulation is so complicated as to horrify the layman. The simplicity of a theory does not depend on how easy it is to solve its equations, but on the crystal-clear conciseness of its principles.

A recurrent theme in the search for increasing simplicity is unification, that is, the discovery that apparently disconnected phenomena are in fact different aspects of the same universal principle. A memorable example is Isaac Newton's remarkable intuition that the fall of an apple and the orbit of the Moon obey the same gravitational law. Another example is the synthesis achieved by James Clerk Maxwell when he identified the very same equations behind electric charges and magnets, revealing the unified pattern of electromagnetism. These were giant steps toward a profound truth: the complexity of nature is underpinned by a unifying scheme and an intangible simplicity.

The journey toward simplicity reached its apotheosis in elementary particle physics, where it has been shown that the structure of matter and fundamental forces descend from a few essential hypotheses, leading to an almost perfect unification of all natural phenomena. We still do not know whether there is a single basic principle, but experimental investigations so far have shown the fundamental physical laws to have a surprising simplicity.

The world around us is complex, though. The mechanisms that generate life, biological organisms, conscious perception, and free will cannot be described with the simple physical laws derived from the world of elementary particles, because they belong to the realm of complexity.

Just as for simplicity, there are different definitions of complexity. Broadly speaking, complexity arises when a system made up of a large number of components displays autonomous characteristics, not recognizable solely in the properties of those components. In short, a complex system is more than the sum of its parts. We will never be able to understand the sprint of a cheetah, the eruption of a volcano, the cry of a distraught man, or the functioning of a computer from a list of all the atoms that compose them. It would be like wanting to understand the meaning of the Mona Lisa's smile from a detailed list of the colors on the canvas. In the system as a whole, there is something that cannot be deduced from its components.

The study of physical reality has taught us that *nature prefers simple principles, but complex manifestations.* In more mathematical terms, the equations that express the fundamental principles of physics are simple, but their solutions, i.e., the ways in which physical phenomena manifest themselves, are generally complex. In more physical terms, the consequences of a physical theory are invariably more complex than the theory itself. In exploring the boundary between simplicity and complexity, we must distinguish between principle and phenomenon. For example, collisions between elementary particles are exceedingly complicated processes, but the principles that govern them reveal a profound logical simplicity.

It is not enough to put together a huge number of components to achieve complexity, i.e., a system that manifests autonomous physical laws. The transition from simplicity to complexity requires far more than a large number of parts. It requires an ingredient that seems almost miraculous, which is hidden in special properties of the physical laws of the microscopic components. For example, by combining Lego bricks, we can build something as large as we wish, but we will never be able to make it complex. On the other hand, by combining elementary particles, nature creates life.

The coexistence of simplicity and complexity is one of the most extraordinary wonders of nature. Complex systems, which shape macroscopic natural phenomena, develop emerging physical laws, in principle deducible from those that govern their components, but in practice completely distinct and capable of acquiring a life of their own. A hidden logical thread links the fundamental laws to those that emerge from them, weaving the complex natural world out of its simple components. The mutual interaction between simplicity and complexity is the ultimate reason for the extraordinary fascination of natural phenomena, whether observed by the enraptured eye of the poet or expressed mathematically in the fundamental principles of the scientist.

Nature has achieved what no artist has ever managed to do. In its masterpiece, it has married simplicity and complexity, two concepts that seem irreconcilable, and yet merge harmoniously to enchant us with a universe where basic fundamental laws and complex physical phenomena coexist.

Whether it is simplicity or complexity that dominates depends on the distance scale at which nature is observed, as shown schematically in Fig. 13.1. The LHC, the large CERN accelerator, scrutinizes the structure of matter down to fractions of billionths of billionths of a meter. This is the realm of elementary particles, where we have derived the most fundamental physical laws known today, those which manifest the extraordinary simplicity of the logical structure of nature. As distances increase, we encounter atomic nuclei at one millionth of a billionth of a meter, then atoms and molecules at a billionth of a meter, cells between ten and a hundred millionths of a meter, and biological organisms on the scale of a meter. At each subsequent step, complexity tends to gain the upper hand over simplicity. New laws emerge and the fundamental laws gradually lose their practical utility, swamped by the combinatorics effects arising in complex systems with ever more components. This is why knowledge of the laws of elementary particles will not help you to repair a washing machine or understand why you unexpectedly fall in love.

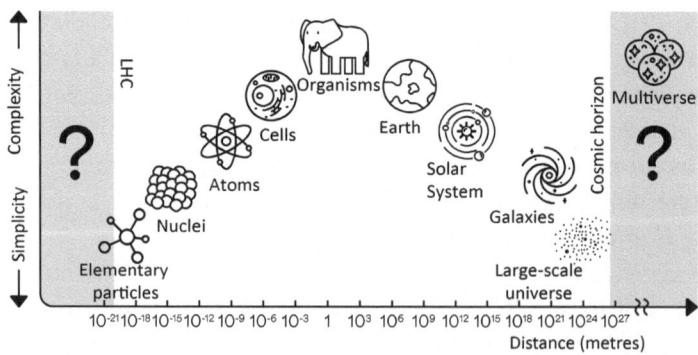

Fig. 13.1 The growth of complexity from the microscopic world of elementary particles to that of biological organisms, and the subsequent decrease in complexity, when we consider the universe at increasingly larger distance scales, up to the limit of our cosmic horizon. The existence of the multiverse would indicate a reversal of this trend at even greater distances. The gray areas in the figure indicate scales we cannot yet explore experimentally

Does complexity increase endlessly with increasing distance? Astronomy says it does not. Exploration at ever greater distances has revealed to us that, after a certain point, the steps change direction. Beyond the planets, stars, and galaxies, at distances greater than a million billion billion meters, simplicity regains the upper hand over complexity. In this regime, the same fundamental physical laws that govern the simplicity of the microscopic world begin to reassert themselves as the rule of law in cosmic space. This emerging simplicity is the deep reason that explains the connection between the microworld and the cosmos. This is the secret that enables us to study the origin of the universe with powerful particle accelerators inside underground tunnels.

Is this the end of the story, with simplicity surfacing at the two opposite extremes of tiny and immense space, while complexity prevails in the intermediate regime? According to eternal inflation, it is not. Continuing beyond the horizon of the observable universe, beyond everything we can measure with our instruments, complexity makes a final flourish and returns as the master of physical reality. The multiverse signals the reappearance of complexity in the furthest regions of space, beyond the boundaries of the cosmic horizon. Almost as in a living organism, the cosmos manifests a new and still unknown form of complexity at the largest distance scales we can conceive of today.

Apologia for Imperfection

The re-emergence of complexity in the multiverse rewrites the research program for a hypothetical final theory. Expecting to find the final answer to the ordering of nature in the physical laws of our particular island universe may be as illusory as trying to deduce the shape of a beach from the study of a single grain of sand.

The lesson we learn from the multiverse is that, even if there were a simple final theory, it could manifest itself in a complex way. There is nothing strange about this, because we have seen over and over again from the study of the physical world that nature prefers simple principles, but complex manifestations. The real novelty of the multiverse is that the complexity is not only manifested in phenomena, but also in physical parameters. As a result, even the physical laws of individual island universes become part of the complexity, as they too become the expression of a phenomenon.

The notion of a pure and crystalline universe, like the almost perfectly flat and uniform one that emerges from astronomical observations, seems to clash with the intuition that nature prefers to manifest itself in a complex way. It also seems to go against the spirit of the Copernican Revolution, which shies away from absolute and perfect cosmic structures. I find much more appealing the image of a universe in perpetual turmoil, where each island universe is nothing special, just one of a population of individuals that are born and age, where the Big Bang is not a fateful moment, just one among infinitely many other episodes characterizing cosmic history, where there is nothing absolute and perfect, just a maelstrom of space in continuous change. To me, the multiverse appears much more in line with what we have learned so far about nature.

Here lies the true conceptual revolution of the multiverse: the idea that complexity could play an essential role in the investigation of first principles. Chasing after some final equation which summarizes all knowledge of nature at its deepest level may just be futile. The search for the key to natural order is a fascinating story, but one of those stories that may never have an end. There may be no absolute truth, only ever better approximations of the truth.

If every universe island is indeed characterized by distinct physical parameters, the search for the principles of nature will acquire probabilistic overtones. From this point of view, physicists exploring the multiverse today find themselves in a similar situation to those at the beginning of the twentieth century, when quantum mechanics shook the scientific world by demonstrating that nature obeys probabilistic laws. It was a traumatic experience for physicists of the time to abandon the deterministic conception of reality which had been taken for granted since the birth of science.

There is, however, a substantial difference. In the case of quantum mechanics, the probabilistic nature of the theoretical prediction can be tested by repeating the experiment an unlimited number of times and collecting a statistical sample of an arbitrarily large data set. In contrast, probabilistic predictions about the multiverse can be compared with only one universe: the one in which we live. This horribly complicates the lives of those who want to explore the multiverse, because it is not clear how to navigate the labyrinthine complexity of its probabilistic predictions.

The Anthropic Principle

To understand one of the logical traps to which the explorers of the multiverse are exposed, imagine you are a fervent player of Go. You are so passionate about the game of Go that you decide to study its popularity in the world. Having a penchant for science, you decide to start from statistical data and conduct a survey among the members of the local Friends of Go Society. The results are surprising. They show that Go is the favorite game of all the people answering the survey. The data are astonishing and you are convinced that you should conduct an in-depth study to understand this unexpected popularity of Go around the world.

There is no need to be an expert in statistics to understand that you are making a mistake, because your data sample is very heavily biased. Statistical considerations about the universe can fall into the same logical trap. Johannes Kepler was a victim of one of these traps in 1596. He believed that the distances to the planets in the Solar System must hide a deep principle of nature. Inspired by a metaphysical concept of cosmic harmony, he hypothesized that the orbits of the six planets then known were contained in a nested sequence of the five Platonic solids, taken to embody the perfection of geometric forms. On the basis of this logical device, he deduced the ratios between the interplanetary distances, obtaining results in good agreement with the measurements of the day.

Kepler's idea was perfectly justifiable. At the time, the Solar System was considered to be the whole universe so it was entirely reasonable to look for some ordering principle of nature in its structure. Today, we know that Kepler made a mistake. The Solar System is just one among an immense multitude of stars with planets orbiting around them, so no universal message can be read into the positions of Jupiter or Saturn. The interplanetary distances are determined by fortuitous circumstances. They conceal nothing fundamental.

A similar logical trap arises in the multiverse when inhabitants of a particular island universe try to deduce the probability of the physical characteristics of their own world. The universe in which we live is extremely sensitive to variations in the physical parameters. If the gravitational force were only a little stronger, stars would burn so quickly that there would be no time for biological evolution to produce complex organisms. If it were only a little weaker, matter would not be able to condense into stable galactic structures. If the electromagnetic force were a little stronger, the repulsion between atomic nuclei would prevent the thermonuclear processes that make stars shine. If the weak force were a little stronger, most atomic nuclei would become unstable and hydrogen would be the only chemically stable element in nature, making the existence of any living organism impossible.

In short, as soon as the physical parameters are modified, even by a very small amount, the universe changes so radically that it becomes incapable of hosting any form of life, because the basic conditions for any complex structure will be lacking. This means that the vast majority of island universes are probably sterile, given that the formation of complex structures requires a delicate balance between the various physical parameters. Our universe is evidently compatible with the existence of complex structures and this means it is rather special among all those that make up the multiverse. Any statistical consideration based on properties of our universe will be heavily biased, just like a survey among the local members of the Friends of Go Society.

To avoid the logical traps related to the specificities of our own universe, we resort to what in physics is known as the *anthropic principle*. Essentially, the anthropic principle tells us that, in evaluating the probability of our own universe within the multiverse, we should not compare it with all the universes allowed by the theory, but only with those that have characteristics compatible with the formation of complex structures. Universes devoid of such structures could not host creatures contemplating the probability of the existence of the universe.

For example, in the case of the problem faced by Kepler, the anthropic principle provides a justification for the Sun–Earth distance on a probabilistic basis, using the existence of terrestrial life as an observational datum. In a universe where planets can form at any distance from their associated star, it is logical to expect the Sun–Earth distance to fall within the range where water can exist in a liquid state on the planet, as we actually observe. If this were not the case, our kind of water-based living organism would not have been able to evolve and we would not be here to reflect on the origin of the Big Bang.

The anthropic principle comes into play in probabilistic considerations where the observer is part of what is observed. It is a logical lifebuoy that prevents us from drowning in a sea of unlicensed conclusions about phenomena we ourselves belong to, or that discourages us from seeking deep explanations in the wrong place.

Scientists ask questions in the belief that their answers will reveal important information about nature. From this point of view, the anthropic principle does not provide a good scientific answer, capable of revealing new truths. Its role as a logical lifebuoy is only to provide a warning that there may be no good scientific answer and that the initial question may not hide any deep truth. The anthropic principle is just a way to avoid the mistake of looking for fundamental principles where there are none—the mistake made by Kepler. It is a way to dissuade yourself from studying the reasons why Go is the most practiced sport on Earth.

The anthropic principle, albeit the product of rigorous logic, is sometimes criticized, even by scientists, although the objections often arise from a misunderstanding of its meaning. Perhaps part of this bad reputation can be put down to the name, a poor choice indeed.

The term 'principle' brings to mind the idea of a deep conjecture, but this is not the case. The question is not whether it is right or wrong: the anthropic principle is a tautology and all that matters is to understand when it is appropriate to use it in probabilistic arguments.

The adjective 'anthropic' raises the suspicion that the existence of humanity is the premise for a scientific theory. The exact opposite is true. The anthropic principle uses human existence as an empirical datum, in a context where different outcomes for the universe are admissible. It reflects the most extreme form of the Copernican Revolution. To be anthropocentric would be to discard the idea of the multiverse, arguing that the principles of nature must necessarily be deducible from experimental observations obtained within our cosmic horizon. Nature does not choose its principles to entertain our human thirst for knowledge.

Whatever you call it, the anthropic principle may plausibly play a role in statistical selection within the multiverse, even though it would be naive to think that it can provide the definitive criterion. If we ever manage to find order in the complicated statistics of the multiverse, other, much more fundamental principles will come into play.

A Problem of Infinities

'Infinity' is a word so familiar in common language that, on the face of things, its meaning seems intuitive to us. Notwithstanding, it is easy to be deceived, because infinity is not a number, but a concept. During a lecture in 1924, the mathematician David Hilbert told a story to explain the logical pitfalls that lie behind infinity.

Hilbert's Grand Hotel has an infinite number of rooms. On a summer evening, a traveler arrives and asks for a room for the night. But it is high season, and unfortunately there is not a single room free. At the insistence of the traveler, Hilbert devises a solution. He asks each customer of the Grand Hotel if they would kindly move to the room with the next number up. The occupant of room 1 moves to room 2, that of 2 to 3, and so on. In this way, every guest has a room, but room 1 has become free to accommodate the newly arrived traveler. Problem solved.

Things get complicated when an infinitely long bus arrives, with an infinite number of passengers asking to be accommodated at the Grand Hotel. Hilbert does not lose heart and, offering a thousand apologies, inconveniences the customers again by asking them to move to the room with double the number of the one they occupy. This time, the tenant of room 1 moves to 2, that of 2 to 4, that of 3 to 6, and so on. In this way, no customer has lost their accommodation, but all the odd-numbered rooms are now free to host the infinitely large group of travelers who just arrived.

But there is no peace for Hilbert. Just as night falls, a train arrives at the Grand Hotel carrying an infinite number of buses, each loaded with an infinite number of travelers, and they all want a room. Hilbert wracks his brains and finally finds a way out. The customer occupying room n will have to move to room 2^n. This means that the tenant of room 1 moves to 2, that of 2 to 4, that of 3 to 8, and so on. Then Hilbert asks the nth traveler of the first bus to go to room 3^n, the nth traveler of the second bus to go to room 5^n, and so on, raising ever increasing prime numbers to the power n. In this way, Hilbert manages to accommodate everyone (and indeed he has some rooms left over, because you never know…).

The story continues with an infinite number of trains, carrying an infinite number of buses, loaded with an infinite number of travelers, and with Hilbert always devising new stratagems. But I must stop here, because one could go on talking about infinity forever.

The paradoxical situations encountered in Hilbert's Grand Hotel reappear in the reality of the multiverse, which is teeming with infinities. The probability that a certain island universe is realized in the multiverse is derived from the usual rules of probability theory. The probability of a certain outcome is given by the ratio between the number of favorable cases and the total number of possible cases. For example, the probability that the outcome of a coin toss is heads is given by the ratio between the number of favorable cases (only one) and the total number of possible cases (two, heads or tails), so it is 50%.

Unlike a coin, which has only two faces, the multiverse has an infinite number of island universes, so every probability will be given by the ratio of infinity to infinity. Determining how many infinities there are within infinity is a tricky problem that tends to lead to paradoxical results. To understand the difficulty, it will be useful to make another visit to Hilbert's hotel.

Suppose you arrive at the Grand Hotel in the low season, when there is no shortage of free rooms. Before Hilbert gives you the keys, you wonder what the probability is of getting a room with an even number. It is natural to think that the probability is 50%. Suppose the rooms are arranged as in Fig. 13.2a along the infinite corridor of the Grand Hotel, with each even room opposite an odd one. Then, the number of even rooms is equal to the number of odd rooms, and the probability we seek is 50%, just like in the game of heads or tails.

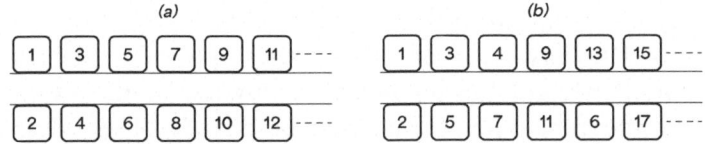

Fig. 13.2 Two possible arrangements of the rooms along the infinite corridor of Hilbert's Grand Hotel. The ordering **a** leads us to think that the probability of being assigned a room with an even number is 50%, since at every step we encounter an equal number of even and odd rooms. The ordering **b** leads us to think that the probability is 25% since, in every group of four nearby rooms (two opposite the other two), only one is even

Walking along the corridor of the Grand Hotel, you realize that the eccentric Hilbert has numbered the rooms as shown in Fig. 13.2b. Although extravagant, Hilbert's choice is perfectly legitimate. In this case, in every group of four contiguous rooms, there is one even room and three odd ones. So, the probability that you get an even numbered room seems to be 25%. But then, is the probability 50 or 25%?

There is no sensible answer to this question. By renumbering the rooms at the Grand Hotel, we can obtain any value for the probability. Each of these values is a fiction, because the ratio between infinity and infinity is ambiguous.

This example serves to illustrate one of the conceptual pitfalls that await us when we seek a statistical interpretation of the multiverse. The probability of encountering certain physical properties could be biased by the way we count the island universes. This issue, known in physics jargon as the *measure problem*, has still not been resolved. Some solutions have been proposed, but it is not yet clear what would be the right criterion for calculating probabilities in the multiverse. But here, we have reached the frontiers of current research and I must stop. The exploration of the multiverse is in its early stages and today we can only hope that future studies will shed light on the many questions that remain to be answered.

Is the Multiverse a Scientific Theory?

Everything that happens in other island universes of the multiverse takes place well beyond our cosmic horizon and therefore remains irremediably invisible, out of reach of any kind of observation. This immediately gives rise to the question: is it legitimate to consider the multiverse as a scientific theory?

The question has sparked a heated epistemological debate. The most common criterion for demarcating the boundary of the scientific method is the principle formulated by Karl Popper, according to which a theory is scientific only if it is experimentally falsifiable. Apparently, then, the multiverse finds itself beyond the boundary of physics, in the swamplands of non-scientific theories, since it can never be directly observed at distances larger than the horizon.

But this way of looking at the question does not stand up to scrutiny. The history of physics is full of ideas that were considered abstract and unfalsifiable at the time, but later proved fundamental for scientific progress. The atomic hypothesis, proposed by Ludwig Boltzmann in the second half of the nineteenth century to explain the properties of gases, was harshly criticized by the positivist currents in physics, perhaps even undermining Boltzmann's mental health and contributing to his suicide. A similar attitude of rejection was reserved for many aspects of quantum mechanics, which were at first considered to lie outside of physical reality. Werner Heisenberg, one of the founders of quantum physics, saw things differently: "The smallest units of matter are not physical objects in the ordinary sense; they are forms, ideas which can be expressed unambiguously only in mathematical language."

Objections to experimental falsifiability were also raised at the time when the theory of inflation was proposed. Some considered it a purely theoretical speculation that could never be put to experimental test, as it described events that occurred in an inaccessible past and were not repeatable in any way. Subsequent research has demolished these objections, and the experimental data we have today test the predictions of the theory to an impressive level of accuracy.

Modern physics operates in extreme regimes—deep within matter, in remote times, or in the immensity of the cosmos—pushing well beyond the frontiers of human perception. No instrument can directly see a quark. The same can be said for the Higgs boson, which lives for no more than a few ten-thousandths of a billionth of a billionth of a second, remaining invisible to any conceivable detector today. These particles live in a reality well beyond the boundaries of direct experimental observability. What is measured are only electronic signals which, through logical deduction, can be interpreted as consequences of the quarks or the Higgs boson. In the physics that explores the frontiers of human knowledge, experimental proof lies not in the observation of the phenomenon, which is usually inaccessible, but in the identification of its consequences.

The situation with the multiverse is not so different. The multiverse is a scientific theory, in the sense that it makes predictions about physical reality, even if this reality concerns humanly inaccessible distances. Unfortunately, unlike quarks or the Higgs boson, current knowledge of the multiverse is too rudimentary to allow us to determine its consequences with any certainty. This difficulty is a human limitation and not a weakness of the theory.

It is undeniable that today the possibility of testing the multiverse theory looks extremely arduous. But human ingenuity has no limits. An experimental comparison could perhaps come from clues left in the cosmic background by hypothetical and fortuitous collisions of our island universe with other wandering islands, or from the collapse of aborted island universes into black holes. Observational research in this direction is underway, but unfortunately the theory offers no guarantee that events of this type will be visible. For the moment, it is more plausible to believe that progress will come on the theoretical side, especially from probabilistic predictions regarding the physical characteristics of our observable universe.

It should be stressed that the cosmic vision proposed by the multiverse is not a hypothesis, but a consequence of the theory of eternal inflation. If we accept a theory, we must also accept its logical consequences. Therefore, the most promising way to confirm or refute the multiverse is to try to consolidate the theoretical edifice on which it is built. Only when we have convinced ourselves that the theoretical structure is sound and reliable will its consequences become an inevitable scientific result. Conversely, if flaws are discovered in the hypotheses of eternal inflation or quantum gravity, scientific interest in the multiverse will quickly wane.

There is still much work ahead of us on the road to understanding the multiverse. The path is anything but simple, but why should we expect understanding the origin of the cosmos to be a simple task? We are in a situation that is not so different from the one described by biologist Thomas Henry Huxley in 1887: "The known is finite, the unknown is infinite; intellectually we find ourselves on a small islet in the midst of an illimitable ocean of inexplicability. Our business in every generation is to reclaim a little more land."

14

Successes and Limitations of Inflation

There is a crack in everything.
That's how the light gets in.
Leonard Cohen

In the last two chapters, we ventured beyond the boundaries of the cosmic horizon. Traveling with the imagination, we were able to glimpse the extraordinary landscape of the multiverse. Now it's time to return inside the observable universe to reflect on what inflation has taught us, and weigh up its successes and limitations.

Cosmic inflation paints a clear portrait of the origin of the universe, based on a simple idea with portentous consequences. The idea is that, before the Big Bang, the universe underwent a phase of furious expansion, fueled by the energy of the vacuum. During this phase, the universe was a cold, dark, and desolate place, where there was only vacuum energy hidden within the folds of space–time. The cosmic horizon remained essentially constant, while space expanded exponentially, escaping beyond the limits of the observable universe. This process of rapid space expansion diluted everything previously existing in the universe; everything, except for the vacuum energy itself, which has the singular property of almost magically reproducing itself, maintaining a constant density, despite the exponentially growing space. Inflation erased the memory of everything that existed before it, preserving only the vacuum energy, which relentlessly powered the overwhelming expansion of space. From this point of view, inflation was without doubt the greatest *damnatio memoriae* ever to occur, capable of eliminating the remnants of the

© The Author(s), under exclusive license to Springer Nature
Switzerland AG 2024
G. F. Giudice, *Before the Big Bang*, Copernicus Books,
https://doi.org/10.1007/978-3-031-69933-7_14

past much more effectively than all the stratagems put in place by the despots that have blighted human history.

The obliteration of everything that is not vacuum energy, as carried out by inflation, may not seem a very constructive way to create a complex universe. But the success of inflation lies precisely in this ruthless stretching out of space, which produces the cosmic circumstances needed to give life to a rich and interesting universe like ours.

The primordial phase of accelerated expansion is capable of explaining those cosmic conditions which, from the point of view of the Big Bang theory, appeared as hypotheses so mysterious as even to seem absurd. But there's more. Inflation identifies those conditions as the only possible outcome of cosmic history. According to inflation, it is inevitable that the universe should find itself, at the instant of the Big Bang, in an expanding space (solution to the mystery of expansion), in a uniform state (solution to the mystery of uniformity), and that the geometry of space should respect the rules laid down by Euclid's axioms (solution to the mystery of flatness). Furthermore, quantum mechanics ensures that there will be density variations, which end up giving life to the galaxies and stars that we observe today in the sky (solution to the mystery of cosmic structure). The enigmas of the Big Bang are explained, one after the other. It is absolutely astonishing how the abstract hypothesis of inflation has, as its only possible consequence, a universe that perfectly resembles the one around us.

It is remarkable to think that the geometry of the space in which we live and the general uniformity of the universe result from the vacuum energy that pervaded space even before matter populated the cosmos. It is even more fascinating to think that all the cosmic structures present in the universe today are the imprints left by quantum fluctuations in the energy of the primordial vacuum and that, without quantum mechanics, the universe would just be a dull, uniform expanse. The physical laws of the microscopic world of particles and the immensity of the cosmos worked together to create the Big Bang, revealing the profound unity of the principles of nature. Inflation teaches us that we live in a quantum universe, where physical reality is the result of a subtle interplay between general relativity and quantum mechanics.

The most amazing aspect of the story is that this intricate understanding of the physical world is not just a fairy tale. It is a scientific hypothesis that can be tested by experimental investigation. Inflation has been brilliantly confirmed by measurements of the flatness of space, the energy density of the universe, the general uniformity of matter and radiation, and the existence of fluctuations that extend beyond the cosmic horizon, with an almost scale-invariant distribution. These data constitute a convincing body of evidence in favor of inflation. Although inflation has not yet been demonstrated according to strict scientific canons, recent experimental investigations have accumulated so many spectacular pieces of evidence to support it that I am convinced of its role in cosmic history.

It is interesting that the theory of inflation provides answers not only about the structure of the universe far back in time (before the Big Bang), but also far away in space (beyond the boundaries of the cosmic horizon). When inflation becomes eternal—still only a hypothesis today—our overall picture of the cosmos changes. Beyond the flat uniformity of our observable universe emerges the multiverse, a multifaceted reality composed of empty space in frantic expansion, dotted with continuously appearing island universes. The idea of the multiverse has completely transformed the search for the fundamental principles of nature, leading us to reflect on the function of complexity in the physical world. It is important to emphasize that, in contrast to the history of the universe before the Big Bang, the picture of the multiverse beyond the cosmic horizon is not currently supported by any experimental data. It is a logical consequence of inflation theory and quantum mechanics, but the multiverse is still only an abstract theoretical conjecture.

Solution to the Mystery of Mysteries

In Chap. 8, I argued that all the mysteries of the Big Bang can be summed up in a single question whose scope is so vast that it reaches beyond the boundaries of science: *was the universe created to host life?* This is the mystery of mysteries.

The initial conditions hypothesized by the Big Bang theory seem perfectly adjusted to lead to human existence. At first glance, it seems unlikely that science could enlighten us on such a point. But the theory of inflation allows us to reconsider the mystery of mysteries, making what seemed an unlikely miracle not only a real possibility, but even an inevitable outcome.

Here lies the wonder of inflation. By flattening the geometry of space, inflation carefully selected the only possible form of the cosmos compatible with human life. By making matter uniform and ruffling it with tiny fluctuations, inflation chose the only combination that would allow biological structures to evolve. Without the dizzying expansion of empty space before the Big Bang, we would not be here today admiring the universe. Inflation offers a scientific explanation of why the universe is so benign with regard to humanity. What at first glance seemed like a supernatural miracle is transformed into a logical consequence.

The result should not be misinterpreted. Inflation does not answer the question of why the universe exists. It only explains why some of its properties do. It states that, if at a certain moment in the past there was a region filled with vacuum energy, the universe must necessarily have the characteristics of geometry and matter required for life to evolve. But inflation does not explain why or how the initial vacuum energy came into existence. In other words, inflation does not offer an ontological explanation of the universe, but pushes the question back in time, well before the Big Bang, to the origin of the inflationary process.

There is another question regarding the mystery of mysteries. The multiverse hypothesis opens a new perspective on the strange coincidences that reveal the exceeding sensitivity of the universe to variations, even very small ones, in certain physical parameters. As discussed in the previous chapter, even minor changes to these parameters would transform the global properties of the universe and destroy the conditions necessary for any form of life to evolve.

The multiverse hypothesis suggests that the enigma may not reveal any deep truth at all. It may just have a simple statistical explanation, based upon the multifaceted variety of island universes. For example, forms of life like our own are based on carbon, which is generated by thermonuclear fusion in the stars. The production of carbon and its subsequent emission into interstellar space require billions of years. So, there is no need to scratch our heads and wonder why our universe has an age of the order of ten billion years, rather than being much younger. If the universe were not so old and vast, we could not be here asking the question. The existence of the multiverse could clarify how some of the mysterious coincidences that predispose the cosmos to the evolution of human life are ultimately nothing more than a banal probabilistic accident.

What Is the Big Bang the Beginning of?

The Big Bang is often portrayed as the initial explosion that gave birth to space, time, and matter. In light of astronomical observations and theoretical knowledge, there is no reason to believe that things really were that way.

The journey back in time, on the wings of Einstein's equation, carries us toward an increasingly dense and hot universe, where spatial distances gradually contract. The onset of a singularity—the moment when space collapses into a point and where the density of matter and temperature become infinite—is only a mathematical property of the equation, not a real physical situation. It signals the existence of a new physical phenomenon which comes into play, changing the regime of the theory and modifying the equations that describe the evolution of the universe. This specific phenomenon is what I called the Big Bang.

The Big Bang was the event that 13.8 billion years ago produced a hot and dense mixture of particles capable of generating the universe as we know it today. This event abruptly changed cosmic history, and the idea of a singularity—that is, a simultaneous beginning of space, time, and matter—is an illusion that arises only if we use the general theory of relativity beyond its limit of validity. A more complete understanding of the Big Bang is required to determine what happened at that moment or even before.

Inflation provides a scientific explanation of the phenomenon that brought about the existence of matter in the universe. The Big Bang was not the explosion of a charge of TNT at a point of space, in the way television shows sometimes suggest. The Big Bang was a uniform transformation in which vacuum energy was converted into thermal energy. It was the moment when particles, up to then trapped in the vacuum substance, suddenly came to life. At this juncture, matter, which was previously in a latent form, hidden within the rigid structure of the vacuum substance, burst forth into a boiling gas containing all possible kinds of particles in thermal agitation. This transformation of vacuum energy into thermal energy almost simultaneously involved a huge, perhaps infinite, space. According to inflation, the Big Bang did not mark the beginning of the expansion of space, but the end of its exponential expansion.

The temperature of the cosmic gas that came into being at the instant of the Big Bang was not infinite. It must have been at least one hundred billion degrees to trigger the thermonuclear processes of nucleosynthesis that produced the light chemical elements in the universe. Moreover, according to inflation, the temperature could not have exceeded one hundred billion billion billion degrees, because otherwise the effect of the resulting primordial gravitational waves would already have been identified in the data of the cosmic background radiation. The search for this effect is one of the most important goals of future experiments. Its discovery would be sensational because it would reveal the temperature of the universe at the instant of the Big Bang.

Inflation allows us to continue the journey back in time, before the moment when the universe was populated with matter. Looking out of the porthole of the spaceship guided by the equations of inflation, we see a cold and empty space, but there is no sign of any catastrophic events, where spatial distances are sucked into a single point. Indeed, from a mathematical point of view, de Sitter's space–time is devoid of singularities. On the journey back in time before the Big Bang, we see spatial distances contracting exponentially, but both the cosmic horizon (which measures the size of the observable universe) and the space–time curvature (which measures the intensity of the gravitational force) remain essentially constant. In other words, inflation does not give any indication that would suggest a Beginning of Everything—the hypothetical event corresponding to the creation of the cosmos. It only shows that the Big Bang cannot coincide with the Beginning of Everything.

Inflation pushes the question of the Beginning of Everything back in time. But by how much? How long did the inflationary phase of the primordial universe last?

We still do not know the answer to this question, but we can at least say this. In order for it to have time to flatten the geometry of space as required by astronomical observations, inflation must have lasted slightly more than 10^{-37} seconds at least: inflation really needs very little time to stretch space adequately. However, it is reasonable to think that inflation lasted longer than the minimum required. For all we know, it could have lasted an enormous length of time, and it is not for nothing that we speak of eternal inflation. In short, even though there are good reasons to believe that an inflationary phase must have occurred in the universe, we know very little about how long it lasted.

What is the Big Bang the beginning of? Why do I insist on calling the Big Bang the beginning of the hot phase of the universe and not reserve this name for the fateful moment of cosmic creation? Wouldn't it be more correct to use 'Big Bang' to refer to the Beginning of Everything? I would argue against, for several reasons.

Firstly, astronomical data indicate that the hot phase of the universe must have undergone a drastic change of regime 13.8 billion years ago, but they say nothing about a hypothetical beginning of physical reality. There is no scientific data that suggests the presence of a Beginning of Everything in cosmic history, let alone that such an event must have occurred 13.8 billion years ago.

Moreover, in order to describe the moment when the universe suddenly became populated with hot and dense matter, the theory of inflation must assume that space existed prior to that. In other words, inflation explains the Big Bang, but has nothing to say about the Beginning of Everything. In short, we do not have reliable clues that point toward a Beginning of Everything, and inflation goes no way toward providing any.

The multiverse helps to illustrate the distinction between the origin of hot matter in the universe and the origin of the global structure of space–time.

Each island universe is born with its own Big Bang, which marks time zero in the cosmic history of that particular universe. However, in the context of the multiverse, Big Bang events happen all the time and none of them indicate the beginning of the entire physical reality. The multiverse theory cannot inform us about the possible existence of a Beginning of Everything. To know more, we would have to go even further back in time and delve into the murky landscapes of quantum gravity.

The Limitations of Inflation

Salvador Dalí said we should not fear perfection, because we will never achieve it anyway. Perhaps the master of surrealism was suggesting that each physical theory represents only a partial description of reality, one that will sooner or later be replaced by a deeper description. Every physical theory will have its limitations and inflation is certainly no exception.

While it solves the mysteries of the Big Bang, inflation raises, in its turn, new mysteries that remain to be resolved. There are two basic issues that reflect our ignorance about the most profound origin of the theory.

Mystery Number 1: What Is the Inflationary Substance Made of?

Inflation hypothesizes the existence of a vacuum substance that permeated the whole of space–time before the Big Bang, but it says nothing about the nature of this substance. The question of its intimate structure, i.e., the particles that constitute it and their role within the scheme of microscopic physics, remains open.

Knowledge of the particle structure would allow us to determine the potential, that is, the relationship between the strength of the vacuum substance and the vacuum energy density. This will be essential if we are to deduce exactly how inflation works and calculate its effects. Astronomical data have already been very important in giving us an approximate idea of the form that this relationship must take, but they will never be sufficient to determine it completely. We will need new information from the frontiers of particle physics.

When inflation was proposed in 1979, the question of finding its particle origin did not seem particularly problematic. The vacuum substance phenomenon is quite common in physics and the zoo of elementary particles seemed rich enough to offer a perfectly adequate choice. The hypothetical particle was given the provisional name of *inflaton*, and it seemed only a matter of time before it would be identified.

Today, the inflaton is still called the inflaton, because no one knows for sure what particle it is. But let's be clear. The inflaton does not have to possess such alien properties that it could have no place in the logical scheme of the microscopic world. Many of the particles that populate the imaginations of theoretical physicists could perfectly well fulfill the role of the inflaton. It's just that none are for the moment entirely convincing.

Particle physicists today find themselves in the situation of a customer who enters a well-stocked milliner's shop in search of a hat to her liking. At first glance, there are so many hats on the shelves that it seems the right one will surely be there somewhere. But, when she tries them on, one is too wide, the other too tight, one clashes with the dress, the other is bland. No matter how hard she looks, none seem really perfect.

Likewise, the problem with the inflaton is that there are infinitely many candidates and yet there are none. There are infinitely many hypothetical particles that fit the bill, but none offer that feeling of fateful inevitability we get when we manage to place the right piece of a puzzle. Identification of the inflaton will be an indispensable step toward understanding and confirming inflation. The hunt is on.

Mystery Number 2: How Did Inflation Begin?

There is a theorem according to which inflationary space–time cannot be extended infinitely far into the past without contradicting the principles of relativity. Put simply, although it can explain the Big Bang, inflation will never be able to explain how the whole of physical reality began. In solving the mysteries of the Big Bang, inflation shifts the question about the origin of the cosmos to much more remote times, but remains silent as regards a definitive answer.

One of the successes of the theory is that, if inflation starts from somewhere in space, its consequences are completely insensitive to the initial conditions. However, we still do not know what determined the initial conditions that triggered the inflationary process and, above all, whether these conditions are achievable in nature. The mystery of the arrow of time tells us that inflation began in a rare configuration of extreme order (low entropy), so cannot arise from a generic state. It must have resulted from a special physical process, whose nature is still shrouded in mystery.

Roger Penrose and other physicists have argued that the initial conditions setting off the inflation process are so unlikely that they negate any interest in its successes. I do not share this inflexible point of view. I see an opportunity, rather than a failure, in the enigma of the initial conditions.

The situation is similar to what happened with the Big Bang theory, which has achieved spectacular successes in explaining cosmic evolution, the background radiation, and the formation of the light chemical elements, although the hypotheses regarding its initial conditions are so special as to seem unreasonable. This was not a good reason to discard the theory. On the contrary, it gave us the key to understanding what lies behind these mysteries. In the same way, the enigma of the initial conditions of inflation is not necessarily evidence of its failure, but perhaps a clue indicating the path to follow to discover what lies behind inflation.

I do not believe that we will ever identify a final theory, capable of answering all our questions about nature. Scientific progress is an endless, patient journey, moving forward step by step. And at each step, while some mysteries are unveiled, others will inevitably be raised, and these will serve to take us further. This was the case for the Big Bang theory and I hope it will be for inflation, too.

The new mysteries we have encountered in this chapter push us even further back in time, to look for what there was before inflation.

15

And What Was There Before?

I would like to tell you what was before the Big Bang.
Unfortunately, there's no time.
Stephen Hawking (apocryphal)

There is a story about a cosmologist who, short of ideas on how to continue his research, went east to ask a wise hermit for insights into the hidden structure of the universe. "The Earth is a flat disk supported by an elephant," replied the sage. Perplexed, the cosmologist retorted: "And what supports the elephant?" "A gigantic turtle," was the sage's response, to which the cosmologist immediately replied: "And what supports the turtle?" Irritated, the sage put an end to the discussion by saying: "Oh, then it's turtles all the way down."

Turtles All the Way Down

Turtles all the way down is a joking expression physicists use to remind themselves of the risk, when looking ever further back in time, of ending up in an infinite regress. Although aware of this risk, physicists don't give up looking for answers to the perennial question of what there was before. What came before inflation? Perhaps a turtle?

The journey back in time in search of what there was before inflation soon becomes arduous, because one quickly encounters conditions under which the universe can no longer be described by general relativity. We enter the

G. F. Giudice, *Before the Big Bang*, Copernicus Books,
https://doi.org/10.1007/978-3-031-69933-7_15

realm of quantum gravity which, as discussed in Chap. 13, is still uncharted territory.

Within this realm, space–time could lose its role as a basic element of physical reality and blur into a shadowy manifestation of some deeper layer of nature. In this case, the very idea of going back in time could lose all meaning, and the search for a Beginning of Everything would just be an ill-posed question. Space–time as we conceive it today may just evaporate into a fuzzy entity as we try to approach a definitive answer to the question of the cosmic origin.

To further the quest for a possible Beginning of Everything, physicists usually choose one of two possible paths. The first is to venture forth into the shadowy lands of quantum gravity, guided by conjectures based on a cock-tail of physical intuition and logical consistency. The other is to take one of the candidates for quantum gravity seriously and study its consequences. In the latter case, string theory is a common choice today, and its rich multidi-mensional structure leaves plenty of room for the imagination of theoretical physicists.

Research on what came before inflation has fueled myriad proposals, bringing in a wealth of new ideas and fascinating phenomena. It is a highly speculative field, largely inaccessible to experimental confirmation. Moreover, the state of theoretical research is still too fragmented and uncertain to lend itself to a synthetic narrative. Just to give a taste of what is at stake, I will present a few ideas that raise stimulating questions, although they are not representative of the whole range of current research. So, this will serve only to whet the appetite of those curious to know what there may have been before inflation.

The Beginning of Time

The beginning of time is a difficult concept to digest. One might ask: what happened a minute before time began? To approach this question, it is better to start from the concept of space. We can travel from Rome to the north, or from Paris to the north, but it no longer makes sense to say this when we reach the North Pole. There is nothing further to the north of the North Pole, even though there is no boundary there. The Earth's surface is a bounded space, but without boundaries.

In the relativistic world, where space and time are merged into space–time, the direction of time could be limited in the past, but without there being a boundary of time. Asking what there was before the beginning of time could be as futile as asking what there is north of the North Pole.

James Hartle and Stephen Hawking explored precisely this possibility. They conjectured that, under the extreme conditions of the cosmic origin, time behaves exactly like space. In this case, there would never be singularities in space–time that correspond to an origin, just as a sphere has a smooth surface, without edges or special points. In this theory, time is limited in the past, but there is no beginning. There is no origin or creation: the universe emerges from nothing.

Hartle and Hawking's theory is based on a conjecture, that is, an assumption without any firm scientific justification. It is the result of an intuition that opens up an imaginary window for us to glimpse certain aspects of quantum gravity.

A different approach to the beginning of time was proposed by Alexander Vilenkin, exploiting a particular feature of quantum mechanics. A quantum state can spontaneously transform into a different state, disconnected from the first, provided that the transition is not prohibited by some physical law. Just as atomic nuclei can suddenly disintegrate through radioactive decays, the geometry of space–time can change as a result of fortuitous quantum transitions. Vilenkin was calculating the probability that a space with spherical geometry undergoes a quantum transition into a de Sitter space, when he noticed a very strange fact. In his calculations, the transition probability remained nonzero in the limit where the radius of curvature of the spherical space became null.

This mathematical result begins to look strange when it is interpreted as physical reality. A spherical space with zero radius is … nothing. It would be like considering the Earth in the limit when its radius is zero: nothing remains. Moreover, to a first approximation, de Sitter space describes the inflationary universe. So Vilenkin had discovered that 'nothing' can suddenly materialize into an inflationary universe without contradicting the laws of physics. This result should be interpreted with caution because the effects of quantum gravity are still unknown and are likely to affect its validity. However, the idea is so intriguing that it deserves some consideration.

Although astonishing, the result may seem only a curiosity for theoretical physicists since, after all, the transition occurs between microscopic quantum states. 'Nothing' transforms into a microscopic de Sitter space and certainly not into a universe as vast as the one that surrounds us. But here the magic of inflation comes into play, greatly enlarging space. Even a microscopic grain of de Sitter space, if left to its cosmic evolution, can generate an enormous universe.

The idea that the matter in the universe might have originated from a quantum fluctuation was considered in the past (and then abandoned). But here we are talking about something much more radical. It is not a quantum transition from an empty space to a space filled with matter. The transition occurs from 'nothing'—that is, a state where space–time does not even exist— to a space–time filled with vacuum energy, potentially capable of creating a universe as vast and complex as ours, and perhaps even a multiverse.

It is hard to conjure up a concrete image of the situation. How should we visualize a physical state in which not even space–time exists? Let's call it 'nothing,' because it's hard to imagine something that is less than the absence of space and time. It would be bordering on the miraculous if this 'nothing' could transform into a real universe. The miracle is a mixture of quantum mechanics, which allows sudden mutations of space, and inflation, which transforms microscopic quantum effects into structures as large as the entire cosmos. Nothing creates the universe.

Inflexible Conservation Laws

The idea of creating space–time from nothing, as suggestive as it is, may seem at first sight inadmissible. In physics, there are incontrovertible conservation laws, and even quantum mechanics must respect them. These conservation laws embody the idea that some physical quantities cannot be created from nothing or disappear into nothing. They determine which changes are allowed and which are forbidden, because the corresponding physical quantities must remain unchanged during the history of the universe. Quantities such as energy, electric charge, and rotational motion do not change in an isolated system. Their total value is conserved.

If we understand nothing as the absence of space–time, all physical quantities must necessarily be zero, because in nothing there is nothing. Conserved physical quantities that are zero at the beginning will remain zero forever. Therefore, if our current universe comes from nothing, all conserved physical quantities must also be zero today. On the other hand, the universe that surrounds us is a system rich in matter, energy, and motion. At first glance, the associated physical quantities do not seem at all to be zero. But upon closer examination, we may be surprised.

In the universe today, electrical charges combine into atoms, so the matter we observe is globally electrically neutral. Astronomical data provides convincing evidence that this is true everywhere in the cosmos. It is therefore reasonable to believe that the total electrical charge of the universe is in fact exactly zero.

All celestial bodies move in space with complicated rotational motions. However, if the universe were rotating as a whole, there would be a privileged direction of space, identified by the axis of rotation. The uniformity of cosmic radiation in every direction is a convincing indication that the universe is not like a spinning top, so the conserved physical quantity associated with rotational motion must be zero.

But what about energy? The universe teems with moving matter and other forms of energy. The energy contributions from matter, radiation, and the vacuum are all positive and therefore, even without calculating their sum, we can be sure that the total is different from zero. However, this computation does not take into account gravitational energy.

The calculation of gravitational energy in general relativity involves subtleties that are too complicated to be described here. However, there are good reasons to believe that the inflationary space from which our universe was born had a total energy exactly equal to zero. The secret is a perfect cancelation between the positive contribution from vacuum energy and the negative one from gravitational energy.

In short, despite appearances to the contrary, there are sound indications to think that, in the current universe, all conserved physical quantities, including energy, are equal to zero. The import of this surprising result is enormous, because it makes it a real possibility that our universe emerged from nothing, that is, a state devoid even of space–time, or at least that it comes from an extremely simple primordial system. Who knows, perhaps nothing is the explanation of everything.

Where Are the Physical Laws Encoded?

The calculation of how the universe can emerge from nothing is based on the probability of a transition between two quantum states. In the initial state, space–time is squashed down to an infinitesimal point, or even disappears into nothing. In the final state, it is a universe undergoing inflationary expansion. For the transition process to have any meaning, the physical laws of quantum mechanics and general relativity must remain valid even for a physical state devoid of space–time. But, can physical laws exist even in the absence of space–time? This is a strange question indeed.

If we start from the Newtonian conception of space and time, understood as non-deformable structures external to physical phenomena, it is intuitive to think that physical laws are part of that rigid architecture. General relativity makes this less obvious, with its malleable and reactive space–time, which plays a dynamical role in physical phenomena and no longer seems a suitable place to host the fundamental laws of nature. So, where are the physical laws encoded? What is the hard disk on which nature stores its laws?

We do not have a good answer to this question, but the exploration of nothing suggests that physical laws have a more structural reality than space–time. The mechanism that produces the universe from nothing may be able to provide us with an explanation for the genesis of space–time, but it remains silent about the origin of the physical laws. Once again, we come back to the eternal question of what there was before. Perhaps another turtle?

The Cosmic Phoenix

If your head starts spinning when you reflect on the beginning of time, perhaps this is the right moment to assume that time has always existed. To do this, we must dust off the concept of an eternal universe, while somehow making it compatible with the cosmic evolution of an expanding space. One compromise between eternity and evolution is to imagine that the universe goes through cyclical phases of expansion and contraction, in an endless cosmic dance.

The possibility of a cyclic world was raised by Friedmann's closed universes, already encountered in Chap. 3, where space is born from the Big Bang and dies in an apocalyptic Big Crunch. A perpetual sequence of bangs and crunches would reconcile the eternity of time with the expansion of space.

The idea was considered as early as the 1930s by American relativist Richard Tolman, and also by Lemaître, the father of the primeval atom, who in 1933 offered an elegant metaphor: "These solutions, where the Universe expanded and contracted in succession [...] had an incontestable poetic charm and recalled the legendary phoenix." But not everyone found the idea so enchanting. In particular, de Sitter declared: "Personally I have, like Eddington, a strong dislike to a periodic universe, but this is a purely personal idiosyncrasy." Indeed, the road to a periodic universe was not without obstacles.

The first problem was to understand how the universe could imitate the mythological Arabian phoenix and rise from its ashes. In more scientific terms, the problem was to identify a physical process capable of initiating a new bang at the end of a crunch. The question was completely out of reach of physicists at the time, because it required them to understand the workings of the Big Bang, and that has only recently been achieved. But, supposing that the universe might mysteriously resurrect itself from the black hole's grave in which it had been buried by the Big Crunch, Tolman came up against an even more general problem.

The second law of thermodynamics states that the entropy of a system (i.e., its degree of disorder) is always increasing. As a result, the universe produced in each cosmic cycle cannot be the same as the one in the previous cycle. The situation is comparable to a snowball rolling across a high-altitude valley, going down one slope, then up the opposite slope, and so on, in periodic oscillations. Rolling up and down the snowy slopes, the ball accumulates snow and hence grows with each cycle. According to Tolman's calculations, the increase in entropy with each cycle lengthens the time between any given bang and the subsequent crunch. This means that, going back in time, the cosmic cycles become increasingly short. So short that we soon encounter a beginning of time, the very obstacle that we were trying to avoid.

Modern string theory has injected new life into the idea of a cyclic universe, because it provides a way to avoid the problem of ever-increasing entropy. There are several interesting proposals in this regard. Each involve complex aspects of string theory and its surprising properties in spaces with more than three dimensions. In the most successful versions, the periodicity does not repeat infinitely often. There is just one contraction phase that bounces back into an expansion phase. Despite various conceptual problems that remain to be sorted out, this line of research is fascinating because it suggests that the beginning of time deduced from Einstein's equation may just be an illusion.

16

The Big Bang Beyond Science

Man is a universe within himself.
Bob Marley

Human inquiry into the origin of the universe goes back to the beginning of civilization. Let us see how our scientific understanding of the Big Bang fits into the history of human thought.

The Voice from Antiquity

Over the centuries, almost every civilization has developed its own creation myth. These are symbolic and highly evocative narratives, relating human understanding to the profound mysteries of the universe and the inescapable laws that hold sway over the flow of events. Myths have allowed humanity to identify its place within the natural order, to give meaning to what seems sacred in the universe, and to relieve the anguish of contemplating the unknown.

A common ingredient in these myths is an exceptional event that determines the origin of the universe—a kind of avatar of the Big Bang. The beginning is often marked by a separation of antithetical elements (the Earth and the sky, light and darkness, ice and fire) that brings about the transition from a primordial state to the natural order of the universe.

For the Maori, the indigenous people of New Zealand, the Sky Father (Ranginui) and the Earth Mother (Papatuanuku) were inseparably united in a reproductive embrace, forcing their children to live in darkness. But some

G. F. Giudice, *Before the Big Bang*, Copernicus Books, https://doi.org/10.1007/978-3-031-69933-7_16

of them, the most reckless and strong-willed, decided to push their parents apart with their vigorous arms and legs. By separating the Earth from the sky, they managed to see the light amidst the pain of Papatuanuku, from whose wounds the gushing blood formed the ochre-colored mountains sacred to the Maori, and the tears of Ranginui, which fall to Earth as rain to show how much he loves her.

There is an echo of the same story in a Babylonian myth in which creation occurs when the god Marduk comes to power, defeating the sea goddess Tiamat by cutting her into two parts that become the Earth and the sky.

These mythical narratives leave open the question of what was there before creation. If a deity created the universe, then who created the progenitor deity? This is similar to the question we asked ourselves when trying to understand what there was before the Big Bang, so it is interesting to see what answers were given in antiquity.

Chaos, a timeless entity without form or origin, is often identified as the primordial element of the universe. In Norse mythology, the primeval chaos was called Ginnungagap, referring to the cosmic abyss, the void. Silence and darkness reigned until the world of ice and frozen rivers (Niflheim) met the realm of fire (Muspelheim), inhabited by giants with flaming swords.

The Taoist tradition also has it that in the beginning there was only chaos, which merged into the primordial egg, a perfect balance between yin and yang, capable of expressing the dualism of opposing but complementary forces. It was the mythical being Pangu who gave life to the universe by breaking the primordial egg with his huge axe and separating yin from yang.

In contrast to a primordial chaos, Christian doctrine supports the idea of *creatio ex nihilo*, which is to say that nothing existed before the origin of the universe. This dogma became dominant in Christian thought at least from the third century on, and was supported by Church Fathers such as Origen of Alexandria and Tertullian of Carthage, but it was only officially accepted in 1215 at the Fourth Lateran Council. Any attempt to find out what there was before creation was not only vain, but even sacrilegious. As Saint Augustine comments in the *Confessions*, to the question: "What was God doing before making heaven and earth?" one could answer with the witty remark: "He was preparing hell for those who pry into such deep mysteries."

The same concept is also found in Judaism. The medieval philosopher and Talmudist Moses Maimonides asserts that the *creatio ex nihilo* is the common theme central to Judaism, Christianity, and Islam. In the Talmudic commentary *Bereshit Rabbah*, probably written between the fourth and sixth centuries, it is noted that the *Genesis* begins with the letter bet (ב) which is closed on three sides and open in the forward direction (given that Hebrew is written from right to left). Its mystical meaning is that we can reflect on what happened after creation, but what happens before creation, above the kingdom of heaven or in the underworld below the Earth, is precluded from human understanding. However, from a philological analysis of the biblical text, it is not clear that the God of Israel created matter from nothing, and even today historians and theologians debate the question of whether the *Genesis* really asserts *creatio ex nihilo*.

A diametrically opposite thesis was advocated by the Greek philosopher Aristotle. Starting from the principle *ex nihilo nihil fit*, according to which nothing can be generated from nothing, Aristotle concluded that the very existence of the universe proves that it must be eternal and immutable. Moreover, the universe cannot be infinite in space, because its natural motion is rotation around the Earth. If it were infinite, there would be points in space that move at infinite speed, a result that was manifestly absurd according to Aristotle.

Between the creation and the immutability of the universe there is a third way. The concept of a cyclic universe, which characterizes Hindu cosmogony, appeared in the *Mahabharata*, an epic Sanskrit poem written in the fourth century BCE. The cosmic cycle of eternal creation and destruction has its counterpart in *samsara*, the succession of life, death, and rebirth in an infinite wheel of reincarnations that is part of almost all the religious doctrines in India. The Indian fascination with large numbers has inspired them to specify the duration of each cycle of the universe, along with a variety of submultiples of the time period, each with its own mystical meaning. Our universe will last a *kalpa*, or a 'day of Brahma,' which corresponds to 4.32 billion years, curiously close to the age of the Earth, today estimated to be about 4.5 billion years.

In summary, the ancient thinking has provided us with roughly four distinct answers to the question of what there was at the beginning: an amorphous and indeterminate primordial chaos, creation from nothing, an eternal and immutable universe, without beginning or end, and a cyclic universe that repeats periodically. It is curious to see how elements of each of these four points of view can be found in the scientific theories examined in this book.

The eternal and immutable Aristotelian universe had a strong influence on science from Newton to Einstein, leading many physicists to resist the idea of the Big Bang. Today we can say with certainty that Aristotle's assumption was wrong. The universe is not at all immutable. It has gone through many upheavals, characterized by phase transitions and profound changes, and will experience others in the future.

Primordial chaos is reminiscent of the physical concept of the vacuum substance and ensuing inflation, while *creatio ex nihilo* brings to mind the idea of space–time emerging from a quantum fluctuation, and the cyclic universe is also a recurring theme in cosmology.

Such resemblances between science and mythology may arouse our curiosity, but they are nothing more than a testament to the boundless resources of human imagination. It also reminds us that scientific progress does not occur in a cultural vacuum. There is always a transfer of ideas, sometimes unconscious, between different areas of human creativity. These fortuitous correspondences have a historical, cultural, and even poetic value, but they should not lead us to believe that there are logical connections between scientific theories, mythical narratives, sacred texts, and philosophical speculations, because each of these areas of human thought operates on a completely different level.

Is There Still Room for a Creator?

Among all the sciences, none encroaches on the field of religion so much as cosmology. Our research on the origin of the universe seems to infringe upon the realm of faith, crossing the boundary between science and spirituality. But this boundary was never absolute. It was always liable to change as human thought progresses. By exploring the logical chain of cause and effect in natural phenomena, science has gradually eroded the territories traditionally associated with religion, always pushing back the boundary between physics and spirituality.

Isaac Newton believed that the gravitational force was a sign of divine intervention. As he declared in his *Principia*: "This most beautiful System of the Sun, Planets and Comets, could only proceed from the counsel and dominion of an intelligent and powerful being." Today, however, we know that gravity results from the deformation of the geometry of space–time.

In 1873, James Clerk Maxwell, discoverer of the unified laws of electromagnetism, stated that a scientific explanation for the origin of molecules would never be found: "We are therefore unable to ascribe either the existence of the molecules or the identity of their properties to any of the causes which we call natural. We have been led, along a strictly scientific path, very near to the point at which Science must stop." For Maxwell, molecules marked the frontier between science and divine intervention. Today, however, we understand the structure and origin of matter at much smaller scales than molecules, in fact down to the level of elementary particles. Once again, the boundary between physics and spirituality has been shifted.

The Big Bang pushes this boundary even further, so it is curious—or perhaps suspicious—that one of its most active proponents was a Catholic priest. Lemaître was well aware of his delicate position within the scientific community. Many of his colleagues were instinctively averse to the idea of a Big Bang because the theory seemed to endorse the account in *Genesis*, concealing some form of Creator. This may also be why Lemaître maintained a strict separation between his scientific activity and his religious faith, while never suffering from any kind of intellectual schizophrenia. Indeed, he even went so far as to say that "the hypothesis of the primeval atom is the antithesis of the supernatural creation of the world." He wanted to make sure that his fellow physicists considered the theory for its scientific content, without experiencing ideological embarrassment.

According to Lemaître, there are two paths to truth. One is scientific and seeks the truth through logical deduction expressible in mathematical equations. The other is spiritual and leads to the salvation of the soul through the revelation of the truth. The two paths are complementary. They respect each other and there is no conflict. In 1933 Lemaître declared: "There are two ways of arriving at the truth. I decided to follow them both. Nothing in my working life, nothing I have ever learned in my studies of either science or religion has caused me to change that opinion. I have no conflict to reconcile. Science has not shaken my faith in religion and religion has never caused me to question the conclusions I reached by scientific methods."

The separation between the two paths to truth means that it is futile to seek evidence of the existence or absence of God by following the scientific method. Lemaître supported the idea of a *Deus absconditus*, inspired by a verse in the *Book of Isaiah*. This is the notion of a God who remains hidden, respecting the autonomy of physical phenomena and human free will. For Lemaître, in the theological sense, *creatio ex nihilo* had nothing to do with the Big Bang, which was a consequence of a phenomenon understandable in terms of physical laws.

Pope Pius XII was less cautious than Lemaître. In his speech *Un'ora*, delivered to the Pontifical Academy of Sciences in 1951, the Pope, who had a great interest in astronomy, hailed the Big Bang theory as scientific evidence in favor of the Christian doctrine of *creatio ex nihilo*. While stressing his conviction that Creation is a revealed truth, lying outside the sphere of the natural sciences, he ventured to say: "Indeed, it seems that the science of today, by going back in one leap millions of centuries, has succeeded in being a witness to that primordial *Fiat Lux*, when, out of nothing, there burst forth along with matter a sea of light and radiation, while the particles of chemical elements split and reunited in millions of galaxies." He even went so far as to criticize the steady-state theory of the universe, although without mentioning it by name.

Fred Hoyle was furious when he heard of the Pope's speech. George Gamow was enormously amused and used it as the basis for jokes of all kinds, arguing that the Big Bang hypothesis now had the seal of papal infallibility. Lemaître, who was present at the speech as a member of the Pontifical Academy and would become its president from 1960, was less amused. He never criticized the Pope publicly but, immediately after the speech, he requested and obtained a private audience. The content of the conversation is unknown, but Pius XII's subsequent public statements on the subject—particularly the inaugural speech at the Assembly of the International Astronomical Union held in Rome in 1952—were certainly much more moderate in presenting the Big Bang in the light of concordism.

Private letters from Lemaître to friends reveal his disappointment regarding *Un'ora*. It is clear that the papal speech exposed the scientist-priest to the criticism from his physicist colleagues that he was pursuing religious ends, despite his firm position to the contrary. Moreover, Lemaître understood that, from a theological point of view, Pius XII was running a serious risk in associating the existence of God with a scientific theory that might well be refuted by data.

At that time, despite Lemaître's stubborn objections, many physicists associated the Big Bang with religious connotations, fueling a general attitude of intolerance toward the scientific hypothesis. A convinced and militant atheist, Hoyle was a leading voice of the general aversion to the Big Bang on purely ideological grounds. Although he never referred to religious implications in his scientific publications, the aim of asserting the thesis of atheism through the theory of the steady-state universe is clearly evident in his popular writings. Hoyle wrote that the concept of the beginning of time is "characteristic of the outlook of primitive peoples," while the hypothesis of the steady-state universe leads to "conclusions for which we might happen to have an emotional preference."

Ironically, in more recent times, the theological connotation of the Big Bang has been reversed. Instead of using it as proof of a divine creation, it is sometimes taken as a justification for atheism. The starting point is the notion that the universe may have originated from a quantum fluctuation amplified by inflation, so cosmic creation occurred from nothing and was determined by physical laws. This removes any need for a creator, which is simply deemed irrelevant. So, this argument would demonstrate the non-existence of God, or at least make God unnecessary.

This has become a standard line of reasoning in favor of atheism, often repeated even by those who understand little or nothing about general relativity or inflation. Alas, the claim is meaningless, from either a theological or a scientific point of view. The weak link in the argument lies in the way the divinity is identified in the context of scientific theory. Space–time is a dynamic entity, like many other quantum fields, and does not play a privileged role in modern physical theories. Rather, the first principle of natural order can be sought in physical laws. This seems to have been the opinion of Einstein, who had a very personal view of God, completely alien to religious dogma: "Certain it is that a conviction, akin to religious feeling, of the rationality or intelligibility of the world lies behind all scientific work of a higher order. [...] This firm belief, a belief bound up with deep feeling, in a superior mind that reveals itself in the world of experience, represents my conception of God."

Einstein identified God with the deep mystery of the existence of a natural order, hence, ultimately, with the existence of fundamental physical laws. As already explained, the creation of the universe from a quantum fluctuation requires there to be physical laws both before and after the transformation of space–time. So, this reasoning does not lead to any theological conclusion.

An alternative is to identify God with the initial conditions of the scientific theory, but even then, we do not get very far, because inflation cannot determine them. To cut a long story short, from a logical point of view, the argument that God can be swept away by the theory of inflation is flawed in every respect.

The mistake lies in resorting to scientific arguments outside their context. Statements about the existence or non-existence of God based on physical theories are profoundly unscientific, because they trivialize the whole idea of the scientific method, which does not apply in the field of metaphysics. Attempting to corroborate theological or atheist theses on the basis of scientific results does a disservice to science, and only ends up damaging its credibility.

A physical theory does not have an ethical, moral, or spiritual dimension, and here lies the boundary with religion. Scientific knowledge neither discourages nor encourages the human desire to reflect on the meaning of life or the value of compassion, leaving such questions to individual choice. Science neither denies nor supports religious thinking, but it can cleanse it of false dogmas, which are often the origin of the atrocities committed in the name of religion, when prejudices and resentments foster violence, racism, and injustice.

Doing science does not require us to adopt any particular position with regard to spiritual matters. Hoyle was a staunch atheist. Gamow was an indifferent agnostic. Einstein had his own original idea of religion. Lemaître was a firm believer. But all these scientists understood each other because they spoke the same language. Science brings us together because it forces us to use a rational language that allows objective comparison and which sticks to mathematical logic and experimental observation as the sole arbiters.

Does Science Make Human Existence Meaningless?

The English poet John Keats accused science of "clipping an Angel's wings" and "unweaving the rainbow." He believed that Newton, by discovering the reflection and refraction of light from water droplets, "destroyed all the poetry of the rainbow, by reducing it to the prismatic colors." For the sensitivity of a romantic poet, science robs nature of its mysteries and stifles imagination with rationality. Even in our time, despite its lack of romanticism, it is legitimate to ask whether physics, by providing a scientific explanation of the Big Bang, removes the enchantment we might feel when contemplating the night sky, just as revealing the magician's trick might take all the magic from the show.

I react in quite the opposite way to Keats. The discovery of the deep mechanisms underlying physical phenomena, rather than emptying them of their beauty, makes us feel the excitement of suddenly seeing nature with different eyes, and of penetrating its most intimate secrets. When we identify the equations that reveal the principles of the universe, the experience surpasses a simple sense of scientific satisfaction and engages the entire emotional sphere. The portrait of the Big Bang painted by inflation is of such vividness that we might almost be present at the spectacle of the origin of matter as it unfolds before our eyes in an extraordinary cosmic display. It would be hard to remain insensitive to such a mind-blowing experience. The more we understand physical laws, the more we come to appreciate the mysteries locked into the order of nature and gain a sense of sharing its deepest secrets.

"The possession of knowledge does not kill the sense of wonder and mystery. There is always more mystery," notes the writer Anaïs Nin in her diaries. Although they refer to the intellectual journey of the artist, her words apply equally well to the scientific understanding of natural phenomena. This is further evidence of the many points of contact between artistic sensitivity and scientific creativity, which, far from being mutually exclusive, actually complement each other.

The more we have learnt about the universe, the more marginal seems the role to which humanity has been relegated. In antiquity, it was believed that the Earth was at the center of a simple planetary system, with the firmament as backdrop. But as our understanding has progressed, the universe has grown in size and diversity, with the discovery of new planetary systems, new galaxies, and an ever-expanding space. Inflation suggests that the universe extends well beyond the limits of the cosmic horizon, while the idea of the multiverse reduces our world to a tiny droplet in an immense ocean.

Is humanity therefore insignificant? Do the vastness of the universe and the inevitability of the physical laws remove all meaning from our existence, abandoning us to an oppressive feeling that we are just dust destined to disappear in a desert of dust?

Of course, we are just a handful of stardust spewed out into space by some supernova. Yet, that handful of dust has managed to understand the universe that surrounds it. Humanity is special because it has consciousness of its own existence, the intellect to ask questions about the origin of the universe, and the desire to search for answers. This awareness helps us to find a meaning and value in human existence. The mystery of the comprehensibility of the universe would remain mute if there was nothing in the universe capable of contemplating it. Human comprehension of the universe gives meaning to the comprehensibility of the universe.

The scientific explanation of the physical world does not destroy the meaning of human existence. On the contrary, it sheds a different light on it, giving us a sharper awareness that we are part of a magnificent framework and providing us with the tools to recognize the deep beauty of the natural order. Understanding the universe instills in us, not a feeling of bewilderment, but one of conscious serenity. It does not convey a sense of conquest, but of connection and complicity with nature.

17

A Story Without End

There is no end. There is no beginning.
There is only the infinite passion of life.
Federico Fellini

Humanity is confined to a tiny corner of space–time. The region occupied by the Earth within the observable universe is smaller than the region occupied by an atom in the Solar System. The duration of human civilization compared to the life of the universe from the Big Bang to today is comparable to a handful of seconds over the course of a year. For the universe as a whole, humanity is just a fleeting sigh.

Yet, from this tiny corner of space–time, humanity has been able to decipher the logical order that governs the universe up to the distances within which we can receive signals, and perhaps even beyond. We have been able to reconstruct cosmic history and discover the existence of an extraordinary event that, by transforming the structure of space–time, gave birth to matter. And we have been able to date this event to 13.8 billion years ago, with astonishing precision for an event that cannot be observed or reproduced, not even in principle.

In about a century, immeasurably greater progress has been made in understanding cosmic history than everything achieved in all the previous millennia. We have moved from a few vague preconceptions about the universe to an accurate theory capable of retracing the cosmic history back to the Big Bang and even beyond, confirming it with very precise astronomical observations.

G. F. Giudice, *Before the Big Bang*, Copernicus Books, https://doi.org/10.1007/978-3-031-69933-7_17

This journey of discovery is the result of a scientific adventure so extreme that it is legitimate to ask: if our brain, our cognitive system, and we ourselves are part of an evolutionary process of natural selection within the universe, how can we transcend the reality in which we are immersed and obtain an independent vision of it? Perhaps, is believing that human beings can deduce the origin of the universe that generated them as vain as believing that a dream can deduce its dreamer? Or that characters in a virtual reality, like the one in the film *Matrix*, can deduce the software that manipulates them?

Quantum mechanics complicates this question even further, because it makes it impossible, even in principle, to unambiguously separate the roles of observer and physical reality. Possibly, our knowledge has reached the point beyond which we can no longer describe the early universe as an external objective reality. Maybe we cannot make further progress in understanding the cosmic origin without a complete theory of quantum gravity, capable of merging general relativity with quantum mechanics.

Even without knowing how to deal with all these issues, cosmology has managed to pursue its exploration back in time toward the origin of the universe thanks to an unexplained feature of the natural world. The mysterious existence of universal physical laws is the ultimate reason why we can find a coherent narration of the entire cosmic history, without having to know all the vicissitudes of every single element of the universe. Herein lies the secret of cosmology, which allows us to describe the universe as if it were a single organism, and not a jumble of disconnected phenomena.

The little girl in the train who observes with intelligent curiosity what the travelers around her are doing is also an essential element in cosmic history. She too is a manifestation of the many phenomena that occur in the universe. However, the purpose of physics is not to describe individual phenomena, but to work back to the most basic principles that govern them. By deducing the fundamental principles of nature and expressing them in mathematical equations, we obtain the key to describing the whole cosmic evolution, from which all phenomena originate—including the little girl. This is the path that has brought humanity so close to understanding the origin of the universe.

The discovery of the Big Bang was one of the most remarkable revolutions in scientific thought. It was born from humanity's insatiable curiosity and nourished by its imaginative ability to combine logical deduction with astonishing experimental observations. I would have no reservations about including the Big Bang among the discoveries that have most contributed to shaping our knowledge of the natural world.

Understanding the meaning of the Big Bang is not just a specialist matter for scientists. It is part of the intellectual identity of every individual in the third millennium. Understanding the Big Bang means becoming aware of the reality of which we are part. It means developing a sense of connection with nature and a sense of belonging to the great scheme that governs the physical laws. It means acquiring the tools to build a personal but conscious awareness of the meaning of existence.

For centuries, humanity has gazed at the night sky with a feeling of awe and wonder. Today, we can do it with an increased admiration for the profound beauty of cosmic harmony, reading in the sky the laws that govern nature from the smallest scales to the most immense.

It sends shivers down the spine to think of the astonishing progress humanity has made in understanding the origin of the universe. Science has propelled us to a level of knowledge unprecedented in the history of civilization, but we should not delude ourselves. The road is still strewn with unsolved enigmas, and the path ahead of us appears boundless. Personally, I do not believe that the narrative of the origin of the universe will ever end with the words 'The end.'

Publisher Correction to: Unveiling the Mysteries of the Big Bang

Publisher Correction to:
Chapter 10 in: G. F. Giudice, *Before the Big Bang*, Copernicus
Books, https://doi.org/10.1007/978-3-031-69933-7_10

Owing to an oversight on the part of the publisher, the original version of the chapter "Unveiling the Mysteries of the Big Bang" was inadvertently published before incorporation of the final corrections. The chapter has been updated retrospectively. A new image and caption was inserted for figure 10.2. The basic facts have not been changed. The publisher apologizes to the author (Gian Francesco Giudice) and his readers.

Additional corrections as in below:

In the Section "The Cosmic Horizon" last three paragraphs are revised at page 108.

The cosmic horizon grows as the universe ages, because the time...................carrying the ships beyond the horizon. What was previously visible, disappears from our sight forever.

Page 109: Fig. 10.2a changed as Fig. 10.2

Page 110: from "we observe today (see Fig. 10.2c). The distance" to "that is reaching us just now (see Fig. 10.2). Today's distance"

Page 110: Fig. 10.2b changed as Fig. 10.2

Page 110: Fig. 10.2a changed as Fig. 10.2

Page 120: Fig. 10.2b,c changed as Fig. 10.2

The updated version of this chapter can be found at
https://doi.org/10.1007/978-3-031-69933-7_10

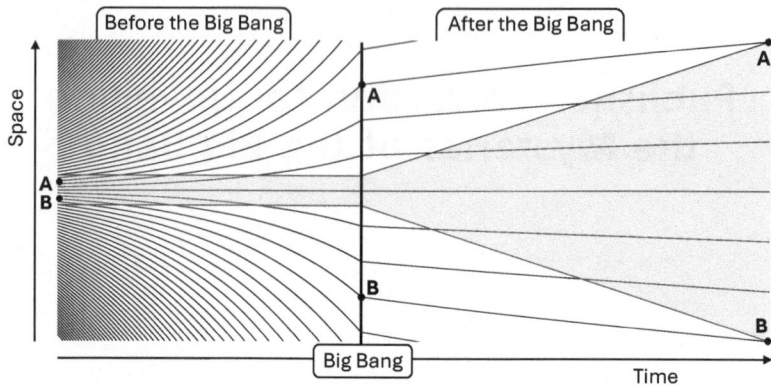

Fig. 10.2 A simplified sketch of the evolution of cosmic distances along one direction of space. The grey area denotes the observable universe, relative to an observer located at the centre. Before the Big Bang, space expands exponentially, escaping out of the cosmic horizon which remains constant. After the Big Bang, the horizon grows faster than space expansion and the observable universe keeps on incorporating new spatial regions. Points A and B, which lie on our current horizon in opposite directions in the sky, have been outside the horizon throughout cosmic history since the Big Bang, but were within the horizon in the distant past, sufficiently before the Big Bang

Epilogue

The two most important days in your life are the
day you are born and the day you find out why.
Mark Twain

I had almost finished writing this book when I went to a theoretical physics institute in South India to give a series of lectures and take part in a conference. The campus was located in an exuberant but carefully managed tropical landscape, with a pleasant balance of beauty and order.

One evening, I went to the campus canteen for dinner. It had been a day of seminars, lectures, and intense discussions with colleagues, and I felt the need to be alone with my thoughts. I queued up with my tray and took my portion of rice and dahl. It was already late and the canteen was half empty. I noticed some people I knew sitting at a table at the back of the room and, to avoid them, I turned furtively toward the garden. Under the trees, there were stone benches and tables for those who wished to eat outside. Almost everyone had already finished dinner. One table was still occupied by a group of students in lively conversation, empty bowls and trays in disarray before them, so I headed straight for a quiet table at the opposite end of the garden.

The evening was mild and the twilight filtering through the leaves of the trees filled the garden with delicate shades of color. The calm surroundings gradually soothed my thoughts. I had almost finished my bowl of dahl, when I noticed someone approaching. On the other side of the table, with tray in hand, stood a young Indian student. She said some kind words about my lectures and then asked if she could sit down to pose a few questions about things she had not understood. Despite myself, I made out it would be no problem and invited her to sit down. She smiled composedly, revealing her

© The Editor(s) (if applicable) and The Author(s), under exclusive license to Springer Nature Switzerland AG 2024
G. F. Giudice, *Before the Big Bang*, Copernicus Books,
https://doi.org/10.1007/978-3-031-69933-7

small and very white teeth, even whiter when contrasted with her long black hair, falling down one side of her face to reach her mouth.

Judging by her age, she must have been at the beginning of her doctoral studies. Her questions were relevant and revealed an excellent understanding of quantum mechanics. As I answered, I waved my hands around as though I were writing on an imaginary blackboard suspended in the air. She took some sheets of paper from her satchel, placed them on the table, and passed me a pen. I reproduced the equations that I had already written down during the lecture, explaining the meaning of each term and giving examples of how to solve them.

Pressed by her questions, I pointed out that my theory was not yet complete, only a conjecture. There were too many aspects I was unable to prove, too many steps left unexplained. My study was based on the hypothesis of the multiverse, which is treacherous territory and hard to navigate.

She liked the topic and began to ask more questions about the multiverse, some of which showed a singular originality of thought. I realized that my answers—inevitably incomplete, since the multiverse still lives in the elusive world of hypotheses—would never fully satisfy her curiosity. As a diversion, I then said that neither I nor the majority of my colleagues would ever be able to solve the issue, because we are trapped within one way of thinking and it takes imagination to break free from conventional patterns. Almost as a joke, I added that perhaps only those who understand Indian philosophy would ever be able to see the cosmos as a single entity, freeing themselves from the cycle of appearances and finding a meaning in the infinities that baffle us today.

After a moment of reflection, the student recited a long passage from the *Upanishads*. Her voice echoed in the now deserted garden and it seemed as if it were no longer hers, but emanated from the branches of the trees above us and from the depths of the Earth, carried by a wind three thousand years old.

When she had finished, the student fell silent and began to eat from her bowl, which had remained full on the table. In the meantime, those verses had transported me elsewhere. They made me think that, even when no one is left to recognize our face or even remember our name, invisible threads of memory will remain. These are the recollections of what we have managed to transmit to those we love and those who come after us. Each of those threads will be there to support and guide future generations, woven into the fabric of values that we call human civilization. Science is part of this fabric too, and here lies the meaning of our research on the origins of the universe.

Suddenly, the student lifted her gaze from the bowl and turned toward me, distracting me from my silent reflections. Her face lit up with a smile, as if she wanted to tell me that she had listened to my thoughts and, in her heart of hearts, agreed with them.

I stared for a long time into her brown eyes, the color of the incense that burns before the statues of Shiva, and then broke the silence: "It's nice to study the universe, isn't it?" She moved her head from side to side in that gesture so common to Indians and so impenetrable to Westerners because, depending on the context, it can mean yes, maybe, or express only the will not to go against one's inevitable destiny.

I inquired about her studies and, when asked if it had been difficult to enter the institute, she confirmed with a sigh. I knew well what that sigh meant. In India, the selection to enter the most prestigious institutes is very strict. Only those who combine exceptional intellectual gifts with unflagging tenacity are admitted. Surely, like many others before her, she had made great sacrifices to indulge her passion for scientific research.

Shadows were falling on the garden, gradually blotting out the features of the girl's face. And the deeper the darkness, the more intense became the scent of India and its mysterious spirituality. We spoke softly so as not to disturb nature as it prepared for the night. When I asked her where she would like to go to do research, she immediately replied: "To CERN," as if she already knew that I would ask her the question. Who knows, maybe one day she really will come to work at CERN and I won't even recognize her.

The darkness was now total and I could no longer distinguish anything around me. From the shadows, I heard her whisper, more to herself than to me: "One day I will be able to answer the question of how the universe began."

Sources of the Quotations

1. p. ix "The universe..." J.B.S. Haldane, *Possible Worlds and Other Essays*, London: Chatto and Windus (1927)
2. p. 1 "I have loved..." S. Williams, *The Old Astronomer*, in *Twilight Hours* (1868)
3. p. 2 "A story..." quoted in F. Gibbons, *Jean-Luc Godard: 'Film is over. What to do?'* The Guardian (12 July 2011)
4. p. 5 "Others have..." P. Picasso, *Metamorphoses of the Human Form*, Munich: Prestel (2000)
5. p. 6 "Absolute, true..." I. Newton, *Philosophiae Naturalis Principia Mathematica, Scholium* (1687)
6. p. 6 "Space tells..." J.A. Wheeler, *Geons, Black Holes, and Quantum Foam*, New York: Norton & Co. (1998)
7. p. 8 "Out of nothing..." J. Bolyai, letter to his father Farkas Bolyai (1823)
8. p. 8 "To praise it..." C.F. Gauss, letter to Farkas Bolyai (6 March 1832)
9. p. 13 "Nothing is..." G. O'Keeffe, quoted in *The Sun* of New York City (5 December 1922)
10. p. 14 "Art is..." P. Picasso, *Picasso Speaks*, The Arts, vol. 3 (1923)
11. p. 16 "of being..." A. Einstein, letter to P. Ehrenfest (4 February 1917)
12. p. 19 "completely full..." E. de Sitter-Suermondt, *Willem de Sitter: Een Menschenleven*, Haarlem: Tjeenk Willink (1948)
13. p. 20 "does not correspond..." A. Einstein, quoted in G.I. Pokrowski, Zeitschrift für Physik 54, 123 (1929)
14. p. 23 "Two roads..." R. Frost, *The Road Not Taken*, in *Mountain Interval*, New York: Henry Holt & Co. (1916)
15. p. 23 "From the one..." E.A. Poe, *Eureka* (1848)
16. p. 24 "pathological personality..." A. Einstein, letter to A. Quinn (1940)

© The Editor(s) (if applicable) and The Author(s), under exclusive license to Springer Nature Switzerland AG 2024
G. F. Giudice, *Before the Big Bang*, Copernicus Books, https://doi.org/10.1007/978-3-031-69933-7

17. p. 24 "to deny the..." W. Nernst, quoted in C.F. von Weizsäcker, *The Relevance of Science*, New York: Harper and Row (1964)

18. p. 25 "I was...," "Well...," "No, I am..." E.A. Tropp, V.Ya. Frenkel, A.D. Chernin, *Alexander A. Friedmann*, Cambridge: Cambridge University Press (1993)

19. p. 27 "Petrograd's honour..." Yu.A. Krutkov, letter to his sister Tatiana (May 1923)

20. p. 28 "On my..." A.A. Friedmann, letter to N.Y. Malinina (1923)

21. p. 28 "I can't..." A.A. Friedmann, letter to N.Y. Malinina (March 1924)

22. p. 29 "The water..." D. Alighieri, *The Divine Comedy*, Paradiso, Canto II, 7 (1321)

23. p. 30 "Some things..." *Revues* of the students of the University of Leuven, 1956–1957

24. p. 30 "beginning of..." G. Lemaître, *Annales de la Société Scientifique de Bruxelles*, 47, 49 (1927)

25. p. 30 "primeval atom..." G. Lemaître, *L'hypothèse de l'atome primitif*, Neuchâtel: Editions du Griffon (1946)

26. p. 31 "Your..." G. Lemaître, *Rencontre avec A. Einstein*, Revue des questions scientifiques, CXXIX, XIX (20 January 1955)

27. p. 32 "I have too..." G. Lemaître, quoted in D. Lambert, *The Atom and the Universe*, Kraków: Copernicus Center Press (2015)

28. p. 33 "It's the best..." Tom Stoppard, *Arcadia*, New York: Samuel French (1993)

29. p. 40 "I did not...," "You have seen..." A.S. Eddington, *Forty Years of Astronomy*, in *Background to Modern Science*, Cambridge: Cambridge University Press (1940)

30. p. 43 "If one..." O. Wilde, *Phrases and Philosophies for the Use of the Young*, Chamaleon (December 1894)

31. p. 45 "This proves...", p. 41 "Apart from..." G. Gamow, *My World Line*, New York: Viking Press (1970)

32. p. 47 "it was a..." R.A. Alpher, R. Herman, *Reflections on Early Work on Big Bang Cosmology*, Physics Today, 24 (August 1988)

33. p. 47 "had never..." quoted in J.D. Watson, *Genes, Girls, and Gamow*, Oxford: Oxford University Press (2003)

34. p. 52 "Gamow was..." E. Teller, *Some Personal Memories of George Gamow*, APS Conference Series, Vol. 129 (1997)

35. p. 53 "There is no..." S. Butler, *Erewhon Revisited*, London: Grant Richards (1901)

36. p. 53 "The idea..." quoted in H. Kragh, *Cosmology and Controversy*, Princeton: Princeton University Press (1996)

37. p. 53 "Philosophically, the..." A.S. Eddington, *The End of the World: from the Standpoint of Mathematical Physics*, Nature 127, 447 (1931)

38. p. 54 "How if..." F. Hoyle, *An Assessment of the Evidence Against the Steady-State Theory* in *Modern Cosmology in Retrospect*, Cambridge: Cambridge University Press (1990)

39. p. 57 "Of course, ..." F. Hoyle, *Home Is Where the Wind Blows*, Mill Valley: University Science Books (1994)

40. p. 59 "To avoid error..." K. Popper, *Objective Knowledge*, Oxford: Oxford University Press (1972)

41. p. 61 "Research is..." W. von Braun, interview in *New York Times* (16 December 1957)

42. p. 61 "I have seen..." A. Haase, G. Landwehr, E. Umbach (ed.), *Röntgen Centennial: X-rays in Natural and Life Sciences*, Singapore: World Scientific (1997)

43. p. 64 "Well, boys..." J. Peebles, AIP Oral History Interviews – Session I (4 April 2002)

44. p. 74 "No army..." V. Hugo, *Histoire d'un crime* (1877)

45. p. 77 "The universe is..." P. de Vries, *Let Me Count the Ways*, Boston: Little, Brown & Co. (1965)

46. p. 77 "In a survey..." Associated Press/GfK survey cited in A.C. Madrigal, *A majority of Americans still aren't sure about the Big Bang*, The Atlantic (21 April 2014)

47. p. 83 "What then is..." Augustine of Hippo, *Confessiones*, Book XI, 14 (398)

48. p. 85 "I confess to you..." *ibid.*, Book XI, 25

49. p. 89 "The people who..." Apple Inc., advertising campaign *"Think different"* (1997)

50. p. 90 "Spectacular..." A.H. Guth, *The Inflationary Universe*, Reading: Perseus Books (1997)

51. p. 92 "The Christian Arab theologian..." John of Damascus, *De Fide Orthodoxa*, Book II Chap. 3 (c. 743)

52. p. 94 "the second..." O. Wilde, interviews for *The New York World* and *The New York Herald* (February 1882)

53. p. 105 "Can we actually..." W. Allen, *Getting Even*, New York: First Vintage Books (1978)

54. p. 110 "a world..." W. Blake, *Auguries of Innocence* (1803)

55. p. 113 "But where..." E.M. Jones, *"Where Is Everybody?" An Account of Fermi's Question*, Los Alamos report LA-10311-MS (1985)

56. p. 113 "Never believe..." attributed to A. Eddington

57. p. 115 "Don't look..." C. Brâncuşi, in P. Pandrea, *Brâncuşi: amintiri şi exegeze*, Bucharest: Meridiane (1976)

58. p. 129 "The only..." S. Dalí, *Diary of a Genius*, New York: Doubleday & Co. (1965)

59. p. 129 *"tantum religio..."* T. Lucretius Carus, *De rerum natura*, Book I v. 101 (1st century BCE)

60. p. 129 "And now, ..." *ibid.*, Book II vv. 1070–1076

61. p. 129 "There are therefore..." G. Bruno, *De l'infinito, universo e mondi*, Third Dialogue (1584)

62. p. 130 "In a letter..." M.T. Cicero, *Epistulae ad Quintum fratrem*, II, 9 (54 BCE)

63. p. 136 " *Maiori forsan* ..." G. Bruno, quoted in a letter from C. Schoppe to K. Rittershausen (17 February 1600). Schoppe, a Lutheran converted to Catholicism, attended the trial by the Inquisition, testified against Bruno, and was present when the sentence was read. In the letter to his friend Rittershausen, a Lutheran who was critical of Catholicism, he describes the trial against Bruno and the popular sentiment in Rome.

64. p. 143 "We adore..." M.C. Escher, note in his diary (4 December 1958) quoted in D. Schattschneider and M. Emmer (ed.), *M.C. Escher's Legacy: A Centennial Celebration*, Berlin: Springer (2003)

65. p. 143 "Life, what..." L. Carroll, *Through the Looking-Glass* (1871)

66. p. 158 "The smallest..." W. Heisenberg, *Das Naturgesetz und die Struktur der Materie*, Stuttgart: Belser (1967)

67. p. 159 "The known..." T.H. Huxley, *On the Reception of the Origin of Species* (1887)

68. p. 161 "There is a crack..." L. Cohen, *Anthem*, in *The Future*, Columbia (1992)

69. p. 176 "These solutions..." G. Lemaître, *L'univers en expansion*, Annales de la Societé scientifique de Bruxelles, A 53, 51 (1933)

70. p. 176 "Personally, I have..." W. de Sitter, Monthly Notices of the Royal Astronomical Society 93, 628, 41 (1933)

71. p. 179 "Man is..." B. Marley, quoted in the Bob Marley Museum, Kingston, Jamaica

72. p. 180 "What was..." Augustine of Hippo, *Confessiones*, Book XI, 12 (398)

73. p. 180 "The medieval philosopher..." Moses Maimonides, *Moreh Nevukhim, The Guide for the Perplexed* (1190)

74. p. 182 "This most beautiful..." I. Newton, *Philosophiae Naturalis Principia Mathematica, Scholium* (1687)

75. p. 182 "We are therefore..." J.C. Maxwell, Nature 8, 437 (1873)

76. p. 183 "The hypothesis..." quoted in D. Lambert, *L'itinéraire spirituel de Georges Lemaître*, Brussels: Lessius (2008)

77. p. 183 "There are..." G. Lemaître, interview in D. Aikman, New York Times Magazine (19 February 1933)

78. p. 183 "Indeed, it seems..." *Un'ora*, Speech by Pius XII to the Pontifical Academy of Sciences (22 November 1951)

79. p. 184 "characteristic of..." F. Hoyle, *The Nature of the Universe*, Oxford: Blackwell (1950)

80. p. 184 "conclusions for..." F. Hoyle, *Frontiers of Astronomy*, New York: Mentor Books (1955)

81. p. 185 "Certain it is..." A. Einstein, *Essays in Science*, New York: Wisdom Library (1934)

82. p. 186 "clipping an…", "unweaving the..." J. Keats, *Lamia* (1820)

83. p. 186 "destroyed all…" J. Keats, during a dinner at the home of B.R. Haydon, on 28 December 1817, with C. Lamb, T. Monkhouse and W. Wordsworth, quoted in B.R. Haydon, *Autobiography and Journals* (1853)

84. p. 186 "The possession..." A. Nin, *The Diary of Anaïs Nin, Volume I* (November 1932), San Diego: Swallow Press (1966)
85. p. 189 "There is no..." F. Fellini (ed. A. Keel and C. Strich), *Fellini on Fellini*, London: Methuen (1976)
86. p. 193 "The two..." attributed to M. Twain, but not documented by historical sources